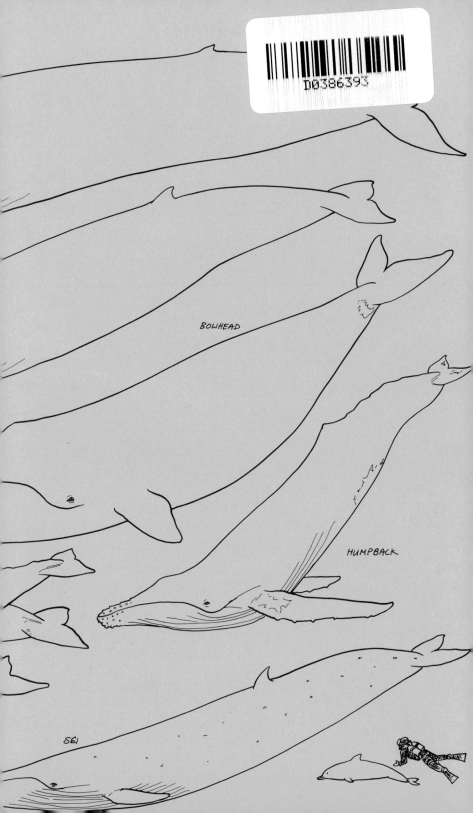

BOWHEAD

HUMPBACK

SEI

THE HUNTING OF THE WHALE

THE HUNTING OF THE WHALE:
A tragedy
that must end

by

JEREMY CHERFAS

THE BODLEY HEAD
LONDON

© Jeremy Cherfas 1988.
Illustrations and diagrams by David Gifford
(© Jeremy Cherfas 1988).

British Library Cataloguing in Publication Data
Cherfas, Jeremy
The hunting of the whale: a tragedy
that must end.
1. Whaling
I. Title
639'.28   SH381

ISBN 0-370-31142-6

Extracts from Lewis Carroll's *The Hunting of the Snark*
are taken from THE ANNOTATED SNARK,
Penguin edition (1967),
edited by Martin Gardner, illustrated by Henry Holiday.

Photoset by Rowland Phototypesetting Ltd
Bury St Edmunds, Suffolk
Printed in Great Britain for
The Bodley Head
32 Bedford Square
London WC1 3EL by
St Edmundsbury Press Ltd
Bury St Edmunds,
Suffolk

For Rachel, with love.

Blubber is blubber, you know; tho' you may get oil out of it, the poetry runs as hard as sap from a frozen maple tree.

Herman Melville, in a letter to Richard Henry Dana Jr,
author of *Two Years Before the Mast*.

Table of Contents

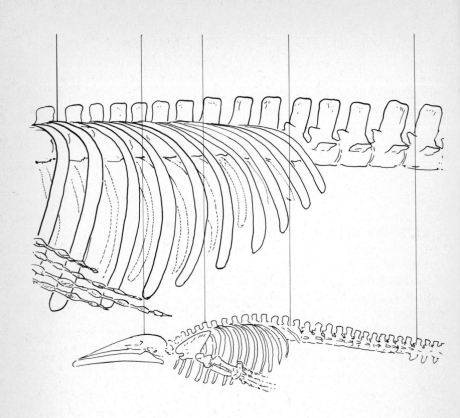

Prologue

The Jury had each formed a different view
    (Long before the indictment was read),
And they all spoke at once, so that none of them knew
    One word that the others had said.

Fit the Sixth: *The Barrister's Dream*

In 1982 the International Whaling Commission voted to ban all commercial whaling for a period of not less than five years. This moratorium was to begin in October 1985. As I write, towards the end of August 1987, at least 11,330 whales have been slaughtered since the start of the moratorium. What went wrong?

The moratorium, when it was passed, was hailed as a great victory. It gave conservationists enormous hope, for they had achieved what ten years earlier had seemed impossible. In 1972 the United Nations Conference on the Environment in Stockholm had voted overwhelmingly for a pause, to take stock of whaling. The whalers, who at the time held sway within the IWC, would have none of it. After a persistent and highly focussed campaign the conservationists managed, a decade later, to gain the upper hand.

Their rallying cry had been "save the whale", and in the first flush of success it seemed to many that they had worked themselves out of that particular job. Shortly after the 1982 meeting, I happened to bump into one of the more effective campaigners on a train. I asked him about his plans for the future. "Whaling's done," he replied. "Tropical forests next." I have not seen him at the IWC since.

It was a glorious campaign. Ordinary people had taken the great whales to their hearts. Much was made of the enormous brain-power of whales, who if only they could speak would surely gain our ear. Instead the campaigners spoke for them. Rationality was sorely missing from the debate, but it was a struggle for hearts and minds and guts, not for brains or reason. Whales were turned into their own

9

best asset. Nobody could fail to side with these magnificent beasts.

There were handy villains too, notably the unfeeling brutes of the Soviet Union and Japan. They were clearly way behind in humankind's glorious advance towards a new consciousness and responsibility. The great whales, exploited to the point of extinction, became a symbol of everything that was wrong with the relationship between people and their planet.

I went to my first meeting of the IWC in 1979, at the Café Royal in London. I was Biology Editor of *New Scientist* magazine at the time, and I wanted to know what all the fuss was about. I believed that whales were a living resource and that the campaign to save them, however noble, was Luddite and misguided. Prudent whalers would want to save the whales themselves. Unlike the campaigners, they were not trying to work themselves out of a job. Since then I have realised that it was I who was misguided.

As it happens 1979 was a pivotal year for the International Whaling Commission, though I do not think anyone realised it at the time. It marked the start of a policy of entryism by whalers and conservationists. The whaling nations encouraged their suppliers to join them in regulating the industry. The conservationists encouraged all manner of countries, some with no historical interest in whaling, to accept that, as the Convention for the Regulation of Whaling has it, "all nations of the world" have an interest in "safeguarding for future generations" the whales. By 1982 the conservationists had the power to defeat the whalers in the IWC; that defeat has singularly failed to carry over into the real world.

A word about the title, *The Hunting of the Whale*. I had been casting about for a title for some time, when in the middle of the night Lewis Carroll's wonderful poem *The Hunting of the Snark* came into my head. Immediately, two or three stanzas sprang to mind, fitting particular chapters perfectly. Although I have tried, I do not have the whole of *The Snark* off by heart, so I got up and read it through. To my delight and relief, there were ten stanzas that might have been written expressly for *The Hunting of the Whale*. That gave me my title, and one stanza to spare. An added bonus was that Lewis Carroll subtitled his poem *An Agony in Eight Fits*, using agony to mean a struggle that involves great anguish, pain, or death. The story of whaling too has been an agony.

I should also explain my system of references. I believe that readers should know the sources of certain passages. In academic biology there is a tradition of citing references as part of the text, a name and date appearing in parentheses after the passage in question. One quickly learns to read over these and ignore or attend to them as appropriate. For people less familiar with this system, I have used footnotes. The footnote on the page refers to the source in abbreviated fashion, and the full details may then be found in the Acknowledgements and References. This, I hope, will enable those who want to chase up sources to do so, while everyone else can ignore them. Anything not referenced in this way is either from a conversation or from a long-since-forgotten source.

This book is a selective history of whales and whaling. It is by no means absolutely exhaustive. Nor is it impersonal. I have skipped over certain aspects, like the long debate on how to kill whales humanely, and dwelt on others, like the struggle over Bowhead whales in the Arctic. I have tried to be accurate, in the facts, although I have also interpreted those facts to give them some meaning. I hope I have been fair, though of course I have probably failed. That is because this is in a sense a work of advocacy.

You might think that none is needed. The battle for whales is won. Perhaps it is, although having studied the problem I am inclined to believe that whaling will end only when the whalers can find no more whales. Certainly that is the historical pattern.

We could choose to escape a future that is a reprise of our past performance. If we do not, that will not mean that all whales are biologically extinct, only commercially. There will be a few left, and those few might one day recover their former numbers. We will have to forget all about them, and our children and their children will have to wait centuries, but it could happen. Unfortunately, animals that exist only in small numbers are very vulnerable to the vicissitudes of life. They are liable to suffer biological extinction as a consequence of disease, or starvation, and ultimately random accident. So even if there are still some whales left when whaling stops, they might not last long.

What does that matter? Life, after all, is just a molecule that somehow achieved the trick of copying itself. Every form of life, human and whale alike, is just another way for the molecule to succeed. Each is an experiment, and each comes to an end eventually. One form has already brought many others to their end. We happen to be the living experiment that has the power to destroy anything it chooses to. I think we ourselves stand a better chance if we actively choose not to exercise our power. Whales offer us that opportunity. If we cannot save the whale, can we save anything?

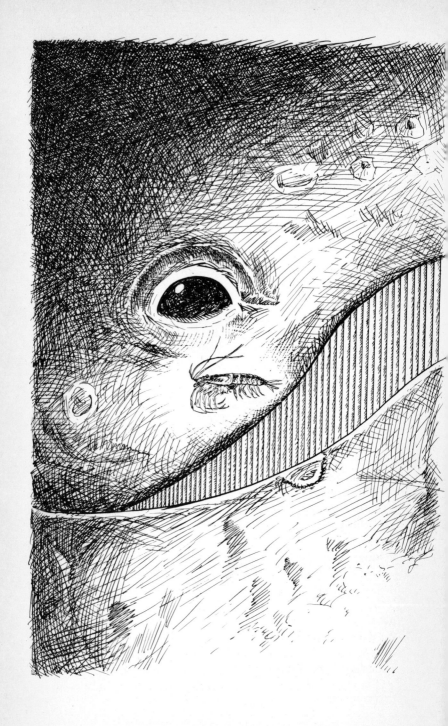

Biology

Come, listen, my men, while I tell you again
The five unmistakable marks
By which you may know, wheresoever you go,
The warranted genuine Snarks.

Fit the Second: *The Bellman's Speech*

Whales, along with dolphins and porpoises, make up the mammalian order known as the Cetacea. There is, in fact, no biological difference between whales, dolphins and porpoises. All are mammals—warm-blooded, air-breathing animals who suckle their young—that have become adapted to a completely aquatic life. Generally the bigger ones are called whales and the smaller ones dolphins or porpoises (dolphins have a pronounced beak, porpoises a blunter head) but size is a bad guide. A far better distinction is between the Odontoceti and the Mysticeti. Odontoceti (from the Greek *odontos*, a tooth, and *ketos*, a sea monster) have teeth while Mysticetes (from the Greek *mystax*, a moustache) have plates of baleen, or whalebone, hanging from the upper palate. Despite their scientific inexactitude, however, the terms whale, dolphin and porpoise are firmly entrenched in common use. Odder still, even though all are mammals their capture is usually referred to as a fishery.

There are ten species of so-called great whales, the ones that have long been commercially exploited. The Southern and Northern Rights and the Bowhead (or Greenland) are Right whales, slow-swimming and easy to deal with once caught; in fact, the right whales to hunt. The Blue, Fin, Humpback, Bryde's, Sei and (taken only since the 1970s) the little Minke are fin whales; they have a dorsal fin, unlike the Right and the Bowhead. They also have many grooves—the better to engulf their food—along the throat. Those pleats earned them their Norwegian name of *royrkval*, or grooved whale, which in English

15

becomes the rorqual.[1] The Gray whale is a baleen whale in a class of its own, while the Sperm whale is a toothed whale.

The naming of whales will have curious repercussions when we come to consider the International Whaling Commission and attempts to regulate whaling; some nations argue that because a whale's name was left off the memorandum drawn up at the time the convention was signed, those whales are exempt from the IWC's deliberations. This is a curious echo of the almost universal folk belief that sorcerers gain their power at least partly by knowing the "true" names of plants, animals, winds, rain and weather. The true names of the great whales, however, show mostly the quite ordinary power of whalers over their prey.

The Right whales—Greenland, Northern, Southern—are obvious enough as they were the right whales to hunt. The Blue and Gray are simple too, for the Blue whale is blue and the Gray whale is grey. Whalers called the Gray whale "devilfish" and "hardhead", commemorating the whale's behaviour. When hunted it often turned on its pursuers and smashed their boats. The Blue is also called the "sulphur bottom", among other names, for its lighter undersides, but these are not really a property of the whale itself. The yellowish colouration is the result of a thin film of diatoms, tiny microscopic plants that grow on the whale's skin. The Blue whale's Latin binomial—*Balaenoptera musculus*—could be a little academic joke. *Balaenoptera* is simply winged whale, a reference to its fins. *Musculus* means well muscled, but might also be a pun on mouse-like, a suitable name for the largest animal that ever lived. Well, taxonomists have a sense of humour too.

One Norwegian whaling captain gained a certain notoriety because he tended to overestimate the length of the whales he had caught. His name was Meincke, and whalers took to calling any small or undersized animals Meincke, or Minke, whales. Eventually the smallest of the great whales came to be known as the Minke whale.

Bryde's whale honours Johan Bryde, the Norwegian consul in South Africa; he built the first two whaling stations there. Bryde's are tropical whales, and were taken off South Africa's

[1] Some people (e.g. Stewart and Leatherwood in Ridgway and Harrison, 1985, p 92) claim "a Norse derivation meaning red whale, referring to the pinkish tint of the many throat grooves when distended".

shores. Sei comes from the Norwegian *seje*, the fish we call pollack or coalfish. Sei whales and *seje* appeared at the same time off the coast of Norway. The whale, however, did not eat the fish; both were feasting on the plankton.

All rorquals have a fin upon their backs, so they are sometimes referred to as finbacks, but the Fin whale's dorsal fin is more prominent than that of other species. Its specific name is *physalus*, a Greek wind instrument, like bagpipes, and also a toad that puffs itself up. A feeding Fin whale, the pleats on its throat distended by a great gulp of food-laden water, does indeed look more than somewhat inflated.

Humpback whales, as they dive, round their backs more than other whales, although they are not really more humped. Their Latin name is *Megaptera novaeangliae*; the giant wing of New England has flippers that can grow to almost one third the length of its body.

Sperm whales contain, in their massive square heads, quantities of spermaceti. This liquid wax, very useful in cosmetics, candles, and many other products, was once believed to be the semen of the whale, hence sperma ceti.

Whales have gone by many names in many different languages. The people who pursued them described their good points, their bad points, or simply their characteristics, Right whales, devilfish, and finbacks. Biologists looking at whales see them from another vantage point, all alike, as whales, but each special as a species.

Right whales (*Balaena glacialis*) are stocky and fat, averaging about 15 metres long. They weigh about 55 tonnes, but their blubber is so thick and full of oil that they float even when dead. That is what made them the correct whales to kill. The body is smooth and round, generally black, with no trace of a dorsal fin on the back. The tail flukes are broad, with pointed tips and a deep central notch, while the flippers are triangular and wide. The head is large, a quarter of the animal's length, with a narrow jaw from which hang the long grey plates of baleen.

Right whales have curious lumps or callosities scattered about the head. These are horny outgrowths which can be 10 centimetres or more high and are usually infested with a selection of parasites: barnacles, worms and whale lice. The callosities tend to be found above the eyes, at the tip of the

17

upper jaw, where they form what whalers and scientists refer to as the bonnet, and in patches along the lower jaw. Most of the callosities have one or more hairs pushing through them. Bearing that in mind (and the very different proportions of the whale's head and our own) the callosities are like the facial hair of human males. The ones above the eyes are eyebrows, the bonnet is a moustache, and the callosities along the lower jaw are a straggly beard. On every whale the callosities look different; Roger Payne, one of the world's leading whale biologists, used the pattern of callosities to identify his subjects in a long study of southern Right whales in Patagonia, and he found that the pattern did not change over the years. Payne has no idea what factors shape the size and pattern of the callosities, but they proved useful nevertheless. The blowholes of the Right whale are far apart on the animal's head, and in still, cold weather produce a V-shaped plume of mist that towers five metres high. Right whales swim slowly, seldom exceeding 10 kph, another characteristic that made them the right whales to hunt. Heavy hunting means that there are few Right whales left, perhaps 200 off the Argentinian coast, a similar number off South Africa, and a total of about 2,000 in the world.

Bowheads (*Balaena mysticetus*) are physically very similar to Right whales, and together they make up the family Balaenidae. Bowheads are, if anything, even stockier than Right whales: the same average length, 15 metres, they weigh much more: 90 tonnes. The head is larger too, up to 40 per cent of the body, and the long, narrow upper jaw arches upward, making room for the extremely long curtains of fine baleen and giving the species its common name. The jaw is so long and so highly arched that two coaches could drive through from opposite sides and not touch. The flippers are rounder than the Right whale's, and unlike the Right whale the Bowhead has no callosities. Bowheads, like Right whales, are very rare, and for the same reasons. The main population haunts the Chuckchi Sea, between the USSR and Alaska, in winter and moves through the Bering Straits into the Beaufort Sea for the summer. This group numbers about 3,000 animals. Other remnant populations are much smaller.

The Gray whale (*Eschritius robusta*) is the only member of the family Eschritidae, intermediate between the Right whales and the rorquals. Females, at 12.8 metres, are slightly larger than the males, which average 12.2 metres. Weights differ accordingly, 31 tonnes for the female, 26 tonnes for the male. The colour is a mottled grey, and instead of a fin there is a series of seven to ten knobs along the mid-line of the lower back. The tail flukes are often worn and tatty, looking distinctly serrated as the whale lifts them prior to diving. Like Right whales, Gray whales have patches of barnacles upon them, but they are not as prominent as callosities and are found all over the body. Gray whales breathe noisily, the blow a sonorous boom that on a still day can be heard more than a kilometre away. The blowholes are close together, and so give rise to a single column of mist. Gray whales often lift their heads vertically out of the water, in a position known as spy-hopping. The assumption is that they are surveying their surroundings, and certainly they seem to be scanning deliberately as they slowly rotate, often taking thirty seconds or more and describing a full circle. Gray whales were very heavily pursued in the nineteenth century, but under strict protection have made a remarkable recovery, and probably number around 15,000 animals in the eastern Pacific.

The remaining species of great whales make up the family Balaenoptera. With the exception of the Humpback, all these rorquals share a strong family resemblance, differing mostly in size and colour.

The Blue whale (*Balaenoptera musculus*) is the biggest, females averaging 26 metres and 105 tonnes. Males are a little smaller, but even so their size is hard to comprehend. An elephant weighs 4 tonnes, so a Blue whale is equivalent to 25 elephants. A cricket pitch, at 22 yards, is considerably shorter than the average Blue whale. A child could easily crawl through the Blue whale's main arteries. Blue whales are long and streamlined, with a small triangular fin set very far back along the body. Beneath the throat are deep grooves, extending beyond the animal's navel. The blowholes are protected behind a kind of splashguard, and the blow is a single strong column that may reach 12 metres above the water. Blue whales are blue, although it is a slaty blue and the belly and tips of the flippers are often lighter. The population has never really recovered from the slaughter of the 1930s, and probably totals about 10,000 animals, 7,000 in the Antarctic and 1,500 in the North Pacific and North Atlantic.

Fin whales (*Balaenoptera physalus*) are very like Blue whales, slightly smaller (21.5 metres, 40 tonnes) and with a distinctive colour pattern. Like all whales, Fins are darker above and lighter below, a pattern known as countershading. On the head of the Fin, this pattern seems to be rotated a quarter turn to the left. The right lower jaw, right baleen plates, and even the right half of the tongue, are pale and unpigmented. Lyall Watson, in his field guide to the whales of the world, suggests that this may have something to do with feeding habits, but does not elaborate. Fin whales are found in all oceans of the world, and are still relatively numerous, perhaps as many as 75,000 animals.

Sei whales (*Balaenoptera borealis*) are smaller than Blues and Fins, and are a steely grey colour. Length averages 15.5 metres and weight about 13.5 tonnes. The fin is larger than that of the Fin and Blue whales, and much further forward on the body.

Next down in size is Bryde's whale (*Balaenoptera edeni*), which seems to be restricted to tropical waters. Bryde's whales average 12.5 metres long and weigh 12 tonnes. They look much like Sei whales, although slightly slenderer and less muscular. Bryde's whale has three ridges along the top of its head, which sets it apart from all the other rorquals, but one seldom gets close enough to see these. Population estimates for Sei and Bryde's whales are confused; they seem relatively high— 80,000 Seis and 20,000 Bryde's—but the whales still need full protection.

Minke whales (*Balaenoptera acutorostrata*) are the smallest of the great whales. Males are only 8 metres long, females an average of 20 centimetres more, and they weigh just 6 tonnes. (The average Blue whale, recall, weighed in at 105 tonnes.) The Minke has a sharp, pointed beak to its head, and looks noticeably squatter than the other rorquals. Colour is generally a dark bluish grey, but there are often white marks on the flippers and a pair of whiter bands behind the head. There are still many Minke whales left, at least in the southern hemisphere, for whalers did not start exploiting them there until recently. Elsewhere, for example in the North Atlantic, where Minkes have been taken for many decades, there are perhaps only a few thousand left.

The Humpback (*Megaptera novaeangliae*) is similar to the other rorquals, but also quite distinct; hence its different species name. This name commemorates its huge flippers, the most notable difference. These giant wings may be five metres long, a third of the body length and almost equivalent to three scuba divers swimming end to end. Humpbacks are also much more thickly built than rorquals, so that although they are only 15 metres long they weigh in at 35 tonnes. The head is long and narrow, and covered with large knobs or tubercles. These are not like the Right whale's callosities, but seem to be fleshy bumps, each with a single coarse hair emerging from its centre. Humpbacks are generally blackish, with white patterns on the fins and the underside of the flukes; these patterns are unique to each whale, and large dossiers of individual whales have been built up from photographs of the flukes taken as the whale begins a dive. There are not many Humpbacks left, perhaps between four and six thousand individuals.

The Sperm whale (*Physeter macrocephalus*) is unlike all the other great whales. It is a toothed whale, and its other common name, the *cachalot*, is a French word derived from the Gascon word for large teeth. Sperm whales are the only great whales in which males are routinely larger than females, 15 metres as compared to 11 metres. They have an unmistakable silhouette, with large, square, blunt-ended head and small, underslung jaw. Only the lower jaw carries teeth, which fit into sockets in the upper jaw and thus maintain a grip on the slippery squid that make up the Sperm whale's diet. The social system too is unique among great whales, a single large bull guarding a harem of twenty to thirty females. The present status of the Sperm whale is one of the more vexing questions that faces the International Whaling Commission; there may be as many as 500,000 animals, but because of their social system, mathematical models predict that the populations might carry on declining even if hunting is stopped completely.

So much for the whales themselves. But as Dean Swift observed, "a flea hath smaller fleas that on him prey; and these have smaller fleas to bite 'em"[2]; whales, the biggest fleas of all, are no exception. Throughout their lives they are besieged by hosts of parasites. Some, like the barnacles, only hitch a ride and probably do the whale no harm. Others, like the whale lice, feed on the whale's flesh.

[2] *Oxford Dictionary of Quotations*, p 528

Perhaps the oddest of an odd lot is the parasite known as *Penella*. This is a Crustacean, a copepod like the copepods in the plankton that sustain so many of the baleen whales. All one normally sees of *Penella* is a thin pink tassel, about 30 centimetres long, hanging from the whale's skin. Carefully cut open the blubber at that point, and you will find something that barely resembles an animal, let alone a copepod. Deep within the blubber is a three-pronged root, *Penella*'s anchor. Between the legs of the tripod is the head, and running from this head to the surface a long, neck-like stem. It is the stem, protruding beyond the skin, that forms the tassel, which carries at its tip the gills and two thin threads used for shedding eggs into the water. This strange tripod is the female *Penella*, feeding on the whale's muscles via long threads that penetrate the blubber. The male is altogether more normal, although tiny and seldom seen by biologists, or anyone else. The body of the animal in the blubber is mostly ovary, and once she has shed her eggs she dies. The wound heals and the eggs, if they are fortunate, find another whale to infest.

Penella is a copepod. So is another animal that calls the whale home. *Balaenophilus unisetus* is an almost microscopic copepod that lives between the baleen plates. It can be so numerous that it forms a whitish scum on the whalebone. It is usually said to do no harm, merely taking advantage of the huge volume of water sluiced past it by the whale, but it seems possible that it might interfere with the whale's own filtering efficiency. (And these have smaller fleas . . . the legs of *Balaenophilus* harbour tiny jug-shaped, single-celled animals, of a genus not yet described by science.)

Coming along for the ride outside the whale are many different animals. One of the largest is the whalesucker *Remora australis*. Remoras commonly attach themselves to larger animals in the sea, and the Cetaceans are no exception. For some reason, the whalesucker prefers Blue whales, and dozens are sometimes seen stuck onto a single whale. The remoras again do no harm, except for spoiling the whale's streamlining a little, because they let go from time to time and go in search of their own food. Not so the Pacific lamprey *Entosphenus tridentatus*, whose sucker and sharp teeth leave a distinctive scar where it has taken a meal from its host. In warmer waters the cookie-cutter shark *Isistius brasiliensis* takes serrated circular mouthfuls of whale skin, leaving characteristic cookie-shaped scars.

The most prominent hitchhikers, none of which do any great damage to the whale, are the barnacles. There are two sorts: the acorn barnacles and the stalked barnacles. Both are highly modified crustaceans which, as Thomas Henry Huxley described them, lie on their backs and kick food into their mouths. Both types are found on whales, although the acorn barnacles seem more common, perhaps because it is easier for their larvae to find a site on the whale to settle. *Xenobalanus globicipitis* is one that colonises almost every species of Cetacean. It is an acorn barnacle that looks as if it is stalked because the shell is very reduced and the body protrudes somewhat from it. This barnacle buries itself partially in the blubber, and lives on the trailing edges of the tail flukes, less often on the flippers or fin. Blue whales often carry several hundred in a solid rank along the edge of the tail, where the animals must withstand tremendous forces when the whale is swimming fast.

Humpbacks are often heavily infested with acorn barnacles, mostly the two species *Coronula diadema* and *Coronula reginae*. They are found on the chin, on the front edge of the flippers, and elsewhere on the body. Often a stalked barnacle, *Conchoderma auritum*, perhaps the most common barnacle in open water, gets a purchase on the shell of some of the acorn barnacles. The Bottlenose whale suffers these barnacles only on its two front teeth, which are always exposed to the water and never come into contact with the upper jaw, but generally the smaller Cetaceans do not have barnacles on them.

Lyall Watson says that at the end of its Arctic summer a single Humpback may be carrying as much as 450 kilograms of barnacle. The other species that carries a great burden of barnacles is the Gray whale. Large clusters of a barnacle known as *Cryptolepas rhachianecti* are found all over the whale's body, but nowhere else in the world. They are unique to Gray whales, which raises the rather interesting problem, not confined to barnacles, of how these animals find their hosts. Whales may be big, but in the vastness of the open ocean they do not present that large a target. Barnacle larvae are induced to settle on surfaces that already carry adults of their species, but it must be an awfully long wait between opportunities to do so, and many larvae presumably go to waste.

Transfer is not such a problem for the other major parasite found on the outside of whales: lice. Reaching down to pat the head of a friendly Gray whale cow in the San Ignacio Lagoon, I was astonished to see hundreds of whale lice, each the size of a small coin. I was even more astonished when one of these voracious carnivores took a bite from my thumb. On one occasion scientists removed more than 110,000 lice from two wounds on a Gray whale, but most infestations are not that heavy.

There are some sixteen species of whale lice, which actually are not lice at all but Crustaceans known as amphipods, related to woodlice and pillbugs. They are equipped with formidable claws on each of their ten legs, and jaws that enable them to take mouthfuls of the whale's skin. Usually they cannot get a grip on the smooth surface of the whale, but wherever there is an irregularity they will be found: among the callosities of Right whales, in the throat grooves of the balaenoptera, tucked into the surface corrugations of the Sperm whale, scuttling among the barnacles on a Humpback, in all the crevices of the blowhole, the angle of the jaw, the anus and genital slit, even on the eyelids. Whale lice have no free-living stage, so they spread only by contact. As one whale is likely to contact only others of its own species, the lice on it are effectively on an island. That is why many of the species of whale lice are unique to particular species of whale. Where one kind of whale louse is found on more than one whale it is a good bet that the whales themselves come into physical contact quite often. *Cyamus monodontis* lives on narwhals *Monodon monoceros* and belugas *Delphinapterus leucas*, which often swim side by side in Arctic waters. Despite these exceptions, the whale's lice can often be used to identify the whale, and as Lyall Watson says, "are a great deal easier to carry back to the laboratory than is the Cetacean itself".[3]

Despite the apparently overwhelming parasite load that many whales carry, these are seldom a cause of death. Indeed, whales have few natural predators, although killer whales *Orca orcinus* will prey on other whales, dolphins and porpoises. If it survives the killer whales, the only other threat to a whale is human activity. In the absence of those threats, some Cetaceans live a very long time. Pelorus Jack was a dolphin who escorted

---

[3] Watson (1980) p 29.

ships into a New Zealand harbour for twenty-four years, and there are other known individuals who were seen again and again for a number of years. But the best information on how long whales live comes from those that died harpooned. It is sometimes possible to tell the age of a whale from a plug that forms in the ear. These plugs have growth rings, rather like trees, one for each year, and thanks to the efforts of whalers we know the maximum ages of many of the species. Most are in the seventies, with Grays at seventy, Bryde's at seventy-two, Seis at seventy-four and Humpback and Sperm both at seventy-seven. The little Minke whale lives to a maximum recorded age of forty-seven years, while the oldest Blue whale captured was 110 years old. The record, however, is held not by a Blue but by a Fin whale; when killed it was 114 years old.

In evolutionary terms, Cetaceans are considerably older than that. Unfortunately, the evolutionary history of whales is not at all clear, so it is not easy to trace the route by which their mammalian ancestors left the land. There are plentiful fossils of ancient whales (Archaeocetes) in rocks laid down 50 million years ago, but these do not look as if they could be the ancestors of modern whales. The Archaeocetes vanish from the fossil record about 20 million years ago, while baleen and toothed whales first appear in rocks a little more than 30 million years old. They are already recognisably either toothed or baleen whales, and so give no hint of the origins of these two groups. The details of the relationships between Archaeocetes, Odontocetes and Mysticetes are still very problematical. The earliest fossil that can definitely be called a whale is a beast known as *Pakicetus*, found in rocks 66 million years old in Pakistan. It seems to have been adapted for a life grubbing around in the shallows. *Pakicetus* probably evolved from animals called Creodonts, primitive ancestors of modern carnivores; the difficult question is whether the toothed and baleen whales themselves evolved from Archaeocetes, or whether they share some joint ancestor among the Creodonts, or indeed whether each of the three groups of Cetaceans is derived independently from a different ancestor.

The latest evidence, based on similarities and differences among various biochemicals, suggests that the Odontocetes and Mysticetes evolved from an ancestor that they share with modern even-toed Ungulates (pigs, cattle, deer, etc.). In fact it

25

seems likely that whales are more closely related to pigs than pigs are to cows. The Odontocetes and Mysticetes shared a long period of common ancestry in the water before finally splitting some time in the Oligocene, about 30 million years ago. The molecular evidence also confirms what anatomy told us: that baleen whales divide naturally into the Right whales and the rorquals, with the Gray whale closely related to the rorquals. It is not yet clear exactly how the various groups of Odontocete relate to one another.

Superficially, all modern Cetaceans are quite similar. They are generally fairly sleek, driving themselves through the water with powerful thrusts of a broad muscular tail. Indeed, powerful is hardly adequate to describe an engine that can, with a few strokes, develop enough impetus to hurl a 40-tonne Humpback completely clear of the water. The horizontal flukes of the tail are not, as some people imagine, the whale's feet. They are simply fins, made of muscle and cartilage. Even the pelvic girdle has almost completely vanished. Just a pair of bones, attached to nothing and embedded in the trunk muscles roughly where the hips might be, remind us that whales' ancestors once walked on land. Of legs there is usually no trace, although whalers sometimes found small mysterious bones in the tail region, and these are almost certainly rudimentary legs. The arms remain, although they have become modified into flippers, which may be broad and spadelike, like a Bowhead whale's, or streamlined ailerons, like a dolphin's, or long flexible wings, like a Humpback's. Beneath the skin there is usually a thick layer of blubber, which keeps the whale warmer in water than a fur coat would. Most of the normal mammalian hair has been lost, although some whales, notably Grays and Humpbacks, do have occasional hair in follicles around the snout.

Cetaceans can be very noisy, and many have the remarkable ability to "see" with their ears. This sense, known as echo-location, depends on emitting loud squeaks and listening to any echoes that return, bounced off things around the sender. The animal then constructs a picture of the environment from the sound information it receives, much as bats do. It seems almost impossible to overemphasise just how astonishing is this ability to echolocate. Dolphins can tell a fish they like from one they don't simply on the basis of echoes. Their ability to discriminate between objects is equivalent to humans being able to tell a small ball-bearing from a grain of wheat at 50 metres. Not all Cetaceans have been tested under the laboratory conditions needed to prove echolocation beyond doubt, but evidence is accumulating that many species can do it, even some of the baleen whales which were previously thought to use sound only to communicate with one another.

The similarities among Cetaceans are, however, partly illusory. They are to a large extent the shared results of life in the water. A streamlined shape, a tail and flippers are all good ideas for moving about in water; seals and sea lions, more recent arrivals in the aquatic environment, show the same changes, albeit to a lesser extent. Water conducts heat very much more effectively than air, so a warm-blooded animal has to expend much more energy on keeping its temperature up if it is in water. Blubber is a very good insulator, keeping the heat in the muscles where it is needed. Whales also have remarkable heat exchangers in the blood vessels that supply the fins and flukes. Arteries carrying warm blood out from the body pass alongside veins bringing cold blood back from the extremities. Heat moves from the hotter to the cooler, keeping the warmth within the body and reducing the amount lost through the fins. Water blocks and bends light, but is very good at transmitting sound and vibration, so the move from vision to hearing is not surprising either.

Biologists recognise two ways in which different animals can resemble one another. The first is a true resemblance; because both species have evolved from a single ancestor, they share some of the characteristics handed down to them from that ancestor. The second kind of resemblance is called convergent evolution; two species that do not share a recent ancestor may be similar because they live in the same sort of environment, and each has solved the problems posed by that environment in a similar way. Birds and bats have converged on turning the arm into a wing to fly through the air. The argument among whale experts is whether the Odontocetes and Mysticetes are similar because a single ancestor took to the water and divided into the two lines once it was aquatic. Or, did two different land mammals return to the water independently, one giving rise to the Odontocetes and the other to the Mysticetes? The balance of opinion is that the two groups are similar because they do indeed share a common aquatic ancestor. But in many ways they are quite different animals.

Perhaps the biggest difference is in the very thing that biologists use to separate them, the teeth: Odontocetes have them, Mysticetes do not. The toothed whales may have a single developed tooth, the "tusk" in the upper jaw of the narwhal, or some 220, as in the spinner dolphin. Generally the teeth are simple conical pegs with a single root. A whale's teeth are usually all the same shape, with no distinction into canines, incisors or molars, and there is only one set. A tooth that is lost or broken cannot be replaced. Some Odontocetes, such as the narwhal and perhaps some of the beaked whales too, probably use their teeth for sparring, although evidence on this is scarce. Those that use them in feeding do not chew with them. Whales swallow their food whole, and the teeth are used simply to grip the food before it is swallowed.

Baleen whales do have teeth, but only as embryos. The teeth are resorbed before the whale is born and their place is taken by the whalebone plates. (This is one of the better pieces of evidence for evolution by natural selection; a perfect Creator, unless she was being very capricious, would surely not design an animal that developed useless teeth only to lose them. They must be a relict of some toothed ancestor. Likewise the two tiny flaps of skin found on every whale embryo less than an inch long: they are an evolutionary reminder of its ancestor's legs.) Baleen is essentially keratin, the same protein that makes hair, hooves, horn and fingernails. The plates develop as outgrowths of the ridges of the upper palate, and they hang down into the jaw. The outer edge is straight, while the inner curves and is edged with a fringe of fine fibres. These fibres, or bristles, act as a strainer for removing items of food from the water.

Each species of baleen whale has its own characteristic size and arrangement of baleen plates. Bowheads have some 350 very long plates with extremely fine bristles. Fin whales have as many plates, but they are shorter and the bristles may be ten times thicker. The baleen whales are a fine example of what ecologists call niche separation. Each species has its own quite distinct bristle size, from the 0.1 mm fibres of the Bowhead to the 1.1 mm of the Blue whale. Thus each species is adapted to filter different-sized organisms from the water. Bowheads take

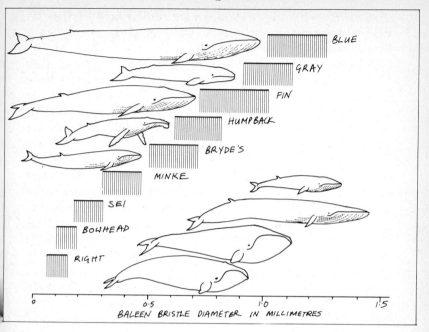

BALEEN BRISTLE DIAMETER IN MILLIMETRES

very fine plankton, while Blue whales specialise in larger items, such as swimming shrimps and crabs. There is almost no overlap in bristle size between different species, and where there is the whales sort themselves out by occupying different parts of the oceans.

All whales use their baleen to strain food from the water, but there are two different techniques for doing so. Right whales tend to be skimmers. They swim slowly along the surface, mouth open, capturing tiny plankton on their fine bristles. To increase their effectiveness they have evolved larger, finer baleen plates. Right whales are almost all mouth.

Finback whales have adopted a different approach. Instead of skimming they gulp at their food, often rushing at it from one side or below. With the jaws open the throat grooves distend to take in a huge volume of water, which is then forced out past the baleen filters. Where Right whales have a large filter area to get more food, Finback whales achieve the same aim by processing a larger volume of water at each attempt.

Gray whales are intermediate, being able to skim and swallow. Sifting through the stomach remains of captured Gray whales has revealed an astonishing diversity of foods, including small fish and plankton and even plant matter such as kelp and eel grass. But most of the Gray whale's food consists of animals living on the bottom, and Grays are unique among whales because they feed on the ocean floor. That has good and bad consequences; bad because they are very seldom caught in the act of feeding, and good because they leave a record of their activity in the mud. The record may be obscure and hard to decode, but it can reveal where the whale has been and what it ate while there.

On one cruise, through the Chirikov Basin of the northern Bering Sea, scientists aboard the research ship *Surveyor* towed a side-scan sonar to look at the bottom. On the traces that rolled out of the machine, covering about 790 kilometres of ocean bottom, they could plainly see patterns, little groups of depressions rather like fingerprints in wet concrete. Sometimes the troughs were in a line, sometimes they formed an arc, and occasionally they radiated out like the petals of a daisy. The depressions measured about two metres long and something over half a metre across, and on average there were about seven prints in the pattern. Each depression was the mouthprint of a Gray whale. The whole pattern represents a single submarine mouthful.

The whale dives down, rolls onto its side, and hovers just above the bottom. Then it pulsates its tongue, sucking up the top few centimetres of ooze and animals. Each depression is one suck of the whale's great tongue. The whale then rises to the surface, often with a plume of mud streaming out behind from its mouth. It may take in some water to sluice the mud from the prey animals trapped by the baleen fringe, and it swallows the mouthful of food before diving down for another.

The sideways roll is an important component of the behaviour, for without it the whale cannot suck up the bottom mud. Most whales have fewer barnacles and more skin abrasions on the right side of the head. The baleen is worn down further on that side too. So most Gray whales are right-headed.

The diet itself is composed primarily of six species of amphipods and isopods, Crustaceans a couple of centimetres long and related to woodlice. Some live in tubes in the mud, and can occur in very dense patches of more than 23,000 in a square metre. They do best in sandy areas that have been physically disturbed, and juveniles prefer to settle down where there are few others. These are just the conditions one finds inside the scoop left by a Gray whale. As a result, you could say that the Gray whale farms its food; the animals the Gray whale eats do best in places where the whale has prepared the ground. Scuba divers found far more juveniles than they expected within the depressions caused by feeding whales.

How much does the whale eat? There are so few records of Gray whales feeding either in the calving lagoons or during the long migration between the nurseries and the feeding grounds that it is safe to assume that almost all of their food is indeed taken in the Bering and Chukchi Seas. The thickness of the blubber does not change much with the seasons, but the amount of oil in it does. During the five months it is on the feeding grounds the average whale gains about 5 tonnes of oil, which represents 15 to 30 per cent of its body weight. Allowing for all the inefficiencies of converting Crustaceans into whale oil, those 5 tonnes need a minimum of 60 tonnes of prey, roughly half a tonne a day.

The figures for Blue whales are even more astonishing. An average Blue whale probably needs about 1.5 million calories a day, but it only feeds for six months of the year, so to sustain itself through the long fast it must take in about 3 million calories a day. Its diet is almost exclusively the Crustacean known as krill, a small shrimp-like animal called *Euphausia superba*. Each shrimp weighs a tenth of a gram and provides less than a tenth of a calorie. In fact a kilogram of krill gives the Blue whale 890 calories, so to get its 3 million calories the whale needs to swallow 3,375 kilograms a day—more than three tonnes. As its stomach holds just under a tonne the Blue whale needs to eat—and digest—four meals a day to stay alive.

Humpback whales, like the other rorquals, sieve food at the surface, but they have also been seen using a feeding technique called bubble netting. The whale (or whales, for sometimes they co-operate) dives down below a school of fish. It then swims in a circle, releasing a steady stream of air from its blowhole. The air rises up as a coiled curtain of bubbles around the shoal. The frightened fish flee from the bubbles and bunch together in a more compact ball. The whale then lunges up through the school, catching far more fish than it would have done without the bubble net.

The plankton that sustains most great whales is found most abundantly in the polar regions. Water cooled by the ice sinks down to the bottom and is replaced by warmer water from below. This upwelling water brings with it sediments and nutrients released from the bodies of dead organisms. In the long, bright summers the combination of sunlight and nutrients permits a fantastic bloom of plant life, the phytoplankton. The plants are mostly single-celled algae, and are not directly accessible to whales in the way that grasses and leaves are available to the largest land animals. They are, however, eaten by tiny Crustaceans, and these animals of the zooplankton support the whales.

There is inefficiency at every step. 1,000 kilograms of plants is converted into about 100 kilograms of zooplankton, which in turn becomes just 10 kilograms of whale meat. Nevertheless, the polar seas, especially the Southern Ocean, are immensely productive. A single hectare near the Antarctic convergence (where the deep waters come to the surface) can produce about 1,200 kilograms of animal protein a year, more than twice as much as the very best pasture on land. The great whales have taken advantage of this bounty, and can fulfil their entire year's nutritional needs in a relatively short time—a Bowhead can skim a year's supply of food in just 130 days.

Perhaps because they are able to feed so efficiently, and perhaps because the food-rich polar waters are so inhospitable for half of the year, great whales make immense journeys of migration each year. The Gray whale feeds off the Alaskan coast, but travels 3,000 miles to mate and calve in the sheltered lagoons of Baja California. Humpbacks journey between Newfoundland and the Caribbean, Hawaii and the North Pacific, and elsewhere. Most of the other rorquals make similar migrations.

During their long migrations whales can sustain speeds of up to 30 kph (16 knots), but top speeds are higher even than this. Killer whales have been clocked at 38 kph (20.5 knots) and Shortfin Pilot whales at 49 kph (26.5 knots). Dolphins regularly manage 45 kph (24 knots) and can probably sprint at more than 56 kph (30 knots). That they are able to do so was once considered little short of miraculous; calculations in the 1930s revealed that the dolphin requires seven times more muscle power than it actually has to move at the speed it undoubtedly does.

The paradox is simply resolved. The calculations are based on the power needed to move a rigid model of a dolphin through the water, but a dolphin is not rigid. The surface of the skin is flexible and undulates as the dolphin swims, thereby reducing the drag of the water. This property is still somewhat mysterious. It cannot be observed in a dead dolphin, even one freshly killed, which suggests that active muscular control of the skin may be involved. There may also be chemical lubricants—oils and other secretions—that help the dolphin's skin to slip through the water. In any case it has proved of enormous interest to the US Navy, which would dearly love to be able to push its submarines through the water as effortlessly as dolphins, and has conducted much research into flexible fuselages for just that reason.

Whales, being mammals, must breathe air. They have developed all sorts of ways to enable them to do so efficiently. The position of the blowhole, at the top of the head, helps them to take a breath as they swim along, but the smaller Cetaceans have perfected another technique; at speed they often leap completely clear of the water to breathe. There are good reasons for doing this. Breaking the surface an animal's body creates maximal drag and turbulence, requiring more energy to maintain speed. By leaping clear it can keep its speed up without wasting too much energy, but to do so it has to breathe quickly. The Fin whale can empty and refill its lungs in two seconds. Humans take twice as long, and the amount of air we shift in that time is 3,000 times less.

The whale's exhalation, the blow so keenly sought by whalers, is simply the vapour in the whale's breath condensing, just as your own breath becomes visible on a cold day. In polar seas, when the air is cold and still, the plume of vapour may hang over the whale for several seconds. In the tropics, it is either invisible or vanishes very quickly. It is not, despite old drawings and stories by whalers themselves, a fountain of water.

Diving—which all Cetaceans do to some extent—poses further problems. Some species submerge for more than an hour, surviving all that time on a single lungful of air. Their lungs are not particularly large, 3 per cent of body weight as compared to 7 per cent in humans, but they are very efficient. We exchange only about an eighth of the air in our lungs at each breath, whereas whales take in up to 90 per cent of fresh air each time they breathe. They have two layers of blood vessels to remove oxygen from the air sacs, as compared to our one, another factor that allows whales to extract half the oxygen in every lungful, whereas we manage only a fifth. Added to this is the enormous capacity that whales have for storing oxygen in their muscles.

The pigment haemoglobin carries oxygen in the blood, and transfers it to a very similar pigment, myoglobin, in the muscles. The myoglobin then supplies oxygen to fuel the power-houses of the working cells. (The "dark meat" of chickens is richer in myoglobin than "white", being those muscles that are used all the time in walking.) Each gram of whale myoglobin can store nine times more oxygen than a gram of human myoglobin. When the whale dives, almost 40 per cent of its total oxygen store is in its muscles, and the meat is so full of the myoglobin pigment that in some species it is almost black.

Going too deep is especially hazardous to human divers. This is because the enormous pressures cause gases, such as nitrogen, to dissolve in the body tissues. Air is only 20 per cent oxygen, the rest is nitrogen and other gases such as carbon dioxide. Normally the nitrogen does no harm, but at high pressure it will dissolve in the bloodstream and also in the fat around the nerves and brain. Divers who surface too rapidly may suffer the bends. The dissolved gases boil off in situ, rather than evaporating in the lungs, and the bubbles thus formed can cause dreadful, even fatal, damage. Sperm whales and bottle-nose whales regularly dive to 3,000 metres for 90 minutes or more. Of course they have only a single lungful of air, as opposed to the tankful a human scuba diver uses up. The nitrogen in a tankful is far more than that in a lungful, and so causes problems when it must be removed from the blood, but even the nitrogen in a whale's single lungful could be enough to cause problems with the bends. Whales avoid absorbing nitrogen by means of an astonishing adaptation. As the whale goes down it simply allows its rib cage, which is less rigid than our own, to fold under the pressure. By 100 metres the lungs have collapsed, completely deflated, and all the remaining air is in the windpipe and nasal passages, which are incapable of absorbing nitrogen. Once that has happened the whale is completely safe and can stay down as long as its oxygen lasts.

Sperm whales have an ingenious device that enables them to hover at any depth they choose without expending any energy. The spermaceti wax that fills the Sperm whale's enormous head may be a focussing lens for the whale's echolocation system, but it is also what divers call an ABLJ, an adjustable buoyancy life jacket. It is a liquid wax, and its density depends on its temperature. So the whale could adjust its buoyancy by controlling the temperature of the spermaceti. The right nasal passage of the Sperm whale is huge, a metre across and 5 metres long, and runs through the spermaceti. It has been suggested that the whale can allow water into this pipe to cool the surrounding spermaceti, thereby controlling its buoyancy at any depth.

How far down do they go? Sperm whales have been found tangled in submarine cables lying on the bottom 1,134 metres down. Sonar operators have tracked Sperm whales down to 1,800 metres, and biologists studying deep dives have records of Sperm whales at 2,500 metres. That is the deepest definite sighting (or, more accurately, hearing, as it relies on detecting the whale's echolocation clicks with an array of underwater microphones). But the circumstantial deep-diving record is held by two large bull Sperm whales killed off Durban in South Africa. They were shot when they surfaced after an eighty-minute dive. In their stomachs were the fresh remains of bottom-dwelling sharks. The water was 3,193 metres deep.

Like all other mammals, whales give birth to live young which they feed on milk. But unlike even the most marine of the other mammals, the seals and sea lions, which come to land to give birth, whales do it all in the water.[4] Because they spend so much of their time far from land and the inquisitive eyes of people, not much is known about the courtship and mating of the great whales. The Gray has probably been better studied than most, and a general pattern of its multiplication is beginning to emerge. A study of the remnants of whales killed as they moved south off the coast of San Francisco revealed that the

---

[4] So do sea otters, giving birth wrapped securely in a "nest" of giant kelp fronds.

embryos came in roughly two distinct sizes. It seems, therefore, that there are two distinct mating periods. One takes place in November, while the whales are moving south, but if the female does not become pregnant then she will produce another egg about forty days later. It is mostly these second matings that scientists have peeped at.

It starts with the female swimming with a motion that can only be described as evasive. She is pursued by males, seven being the most ever seen. One by one males drop out of the chase until there are just two left. The cow slows down, allowing the bulls to catch her up, and as they do so she spirals up to the surface, breaking the water on her back, belly up. The males swim up on top of her, grazing her genital area, which is protected within a muscular slit. Swimming stops, and the female may roll like a fishing boat as the males attempt to mount. They lie on either side of her often clasping her with their fins as they try to steady her and stop her rolling. The weight of the males often causes the female to sink tail first, and as she does so her head often lifts clear of the water. The males pivot with her, all the while keeping as close a contact as possible. She may make a shallow dive, and the males follow, three tails waving above the water. This foreplay is astonishingly gentle, contact usually a caress rather than a crash. Eventually, after about half an hour, it has aroused the female to the point where she simply lies upside down at the surface, her flippers extended. It is hard to avoid the impression that she is, as Californians might have it, blissed out.

The males, also upside down, now swim up on either side of her, each with his pink penis exposed.[5] Each male jockeys in the waves as he tries to make contact with the female's vaginal opening; his curved penis extends more than 1.1 metres, quite long enough to reach over her belly. The penis is muscular too, and the male can to some extent guide it into position. Once contact is achieved the pair roll towards each other to complete the coupling. They may stay joined for up to a minute, the male's penis pumping visibly, but even when they separate that does not denote the end of the affair. The threesome often stays together for at least the rest of that day, and sometimes longer.

[5] The guides on whale-watching tours often joke in an embarrassed fashion about the male's "Pink Floyd".

Perhaps the most remarkable thing about Gray whale copulation, apart from the fact that they ever get it together in the swell and chop, is that the males are not in the least bit belligerent. A single male *can* mate successfully with a female, given enough time, perseverance and good fortune, but a pair will achieve union much more rapidly. Each steadies the female for the other, and they seem content to leave it to luck which of them will actually penetrate her.

The other whales, too, have been observed courting and mating, and the pattern is generally the same: extensive courtship followed by a relatively brief copulation. Charles Scammon, the Californian whaling captain, described the mating of Humpbacks in his book *Marine Mammals of the North-Western Coast of North America*:

> In the mating season they are noted for their amorous antics. When lying in the side of each other the megapteras[6] frequently administer alternate blows with their long fins, which love-pats may, on a still day, be heard at a distance of miles. They also rub each other with these same huge and flexible arms, rolling occasionally from side to side, and indulging in other gambols which can easier be imagined than described.[7]

Whalers believed, probably correctly, that males were spent, and easier to capture, after mating.

The seemingly co-operative mating behaviour of Gray whale males probably conceals seething competition below the surface, within the female. Theoretically, one expects males of all species to compete with one another to sire as many young as possible. Usually the competition is overt, the males fighting to monopolise access to the females. It is precisely because this open competition is the usual state of affairs that the male Gray whale's friendliness to others is so surprising. But there are other ways to compete. The males can allow one another to copulate with the females, but produce more sperm and thus hope to win the race within the female's reproductive tract. If they are going to do this, males will have large testes and a long penis, so that they can deposit more sperm closer to the egg.

[6]Humpbacks.
[7]Quoted by Winn and Winn (1985) p 75.

Sperm competition, as it is called, is a well-known phenomenon, documented in humans and our nearest relatives the gorilla and chimpanzee. Gorillas have tiny testes; the silverback is master of his females and prevents other males from mating. Chimpanzees have enormous testicles and relatively long penes. Their mating system is often uncharitably described as a gang-bang; on many occasions, all the males in a troop will mate with a receptive female within a short time of one another, so the successful father is likely to be the one that produces the most sperm. Humans have rather average testes, and like most mammals we are rather average in our mating system, mildly polygynous. (Our large penis, so spectacularly celebrated by Desmond Morris in *The Naked Ape*, probably owes its size to other influences.) Sperm competition also, apparently, applies to the great whales.

Robert Brownwell, a researcher with the US Fish and Wildlife Service, and Katherine Ralls of the Smithsonian Institution in Washington DC, measured the testes and penes of all the whales they could. Rights, Bowheads, and Gray whales all stood out from the crowd. Their testes were larger than expected; Right whale testes can weigh up to a tonne, six times larger than would be expected from the animal's size alone. Gray whale testes are twice as large as they "should" be. These whales also have relatively long penes, 2.3 metres long in the Right whales, more than 14 per cent of the body length. Gray whales, as their testes would lead us to expect, are not so extreme, with an average penis length of 1.38 metres, 11.5 per cent of body length. Brownwell and Ralls conclude that Right whales, Bowheads and Gray whales, although they seem friendly on the surface, compete within the female.

Humpbacks, by contrast, have relatively small testes. They tip the scales at just 38 kilograms, about half the size they ought to be. And although nobody has made detailed records of Humpbacks courting, it seems from other aspects of their behaviour, notably the song (see pp 50–5), that males do indeed compete directly for access to females. Hence they have no need for massive sperm factories. (Sperm whales did not figure in this analysis, being toothed whales. They operate a true harem system, big bull males defending a group of females against the incursions of other males.)

Twelve to thirteen months after mating, having made a

round trip of 11,000 miles, the Gray whale mother is back in the lagoon to give birth. On the few occasions birth has been witnessed the calf emerges headfirst, and the moment of birth is probably the most dangerous in the whale's natural life. The infant enters a world of water, but must breathe air. The mother often gently supports her newborn, either with her flippers or on her own back, holding its blowholes above water until its regular breathing rhythm has become established.

The newborn Gray whale is about 4.5 metres long, and weighs about 800 kilograms. It is pinkish grey, with quite distinct folds left over from the year it spent curled up in the womb. For a few days the calf is wobbly on its fins, unco-ordinated and erratic, often lifting its whole head above the water to breathe. Quickly, however, it learns how to breathe as easily as its mother, little more than the blowholes breaking the surface. Quickly, too, it grows. Gray whale milk is the richest known, 53 per cent fat, and although nobody knows exactly how fast a calf grows in the lagoons, one that was raised in captivity put on 7 kilograms a day. She was weaned after seven months, by which time she weighed almost 5 tonnes.

Calves spend most of their first year alongside their mothers. They often play, with balls of kelp, with other calves, with their mothers, and even with the skiffs of scientists and sightseers. They also sometimes need rescuing. Steven Swartz and Mary Lou Jones, scientists who founded Cetacean Research Associates Inc. in California, have twice seen adults dramatically come to the aid of a young calf lost in the intricate drainage channels of the San Ignacio Lagoon:

> From our observation tower, less than fifty feet from the deep channel, we saw a baby whale swim over the shallow sand bar toward shore until its belly touched the sloping sand beach. Then, in approximately one meter of water, it began rolling and thrashing. Shortly, two adult whales lunged out of the channel, with heads held high as if to scan the shore ahead, and slid up onto the bar, beaching themselves on either side of the calf. Sandwiching the young whale between them, both adults thrashed their heads and flukes so hard that it rocked their bodies, whereupon they turned by pivot-ing on their bellies and, holding the calf between them, slid back into the deep channel and were gone.

To our amazement, we observed such an event a second time in the same location. Both times the rescue manoeuver was performed within fifteen to twenty seconds, and appeared as a deliberate and coordinated activity on the part of both adults, one of which we presume was the calf's mother. . . . It certainly looked as if the whale assisting the mother knew what was required and how best to offer assistance.[8]

Some time between five and eleven years later, the calf is sexually mature, and begins its own reproductive effort, closing the breeding cycle.

Mammals as a group have a fairly straightforward relationship between the size of the offspring and the duration of its gestation. Horses carry their foals for eleven months, rhinoceroses for eighteen and elephants for twenty-two, all more or less in line with the adult animal's size. Baleen whales have the largest young in the world, but by far the shortest gestation, relative to their size. A Blue whale calf may be almost 8 metres long and weigh 7 tonnes, but it takes less than a year to reach this size. Indeed, none of the baleen whales has a gestation period longer than twelve months. This extraordinarily rapid development is both caused and necessitated by the great whales' diet.

The plankton on which whales feed is incredibly rich, but it forms quite discrete patches in both space and, more importantly, time. Polar plankton blooms prolifically once a year. Females build up their reserves in the polar waters during the summer plankton bloom. Those that are pregnant shunt resources to the developing embryo. As winter comes they migrate to the tropics and more temperate waters to give birth, after which mothers supply their young with prodigious quantities of extremely rich milk. Some of the larger baleen whales may produce up to 600 litres of milk each day. Whale milk contains 40 per cent or more fat, while cows' milk has 5 per cent fat. On this rich diet a calf can double its weight in the first week, and it is not that small to begin with. After two or three months cow and calf set off together on the long migration back to the feeding grounds. Weaning takes place there, so that the

[8] Swartz and Jones (1984) p 18.

young whales can eat as much as they need. The mother, meanwhile, continues to replenish her reserves for the journey back to the tropics, where she may be mated again.

Not all whales follow this pattern. Sperm whales, unusual in so many respects, gestate their young for between fourteen and sixteen months. They do move towards the poles to feed, but not as far as the baleen whales. The males migrate further and tend to leave the females behind in more tropical waters. The calves are thus born shortly after the males have departed; the unruly breeding harems, full of squabbling males, settle down into quieter nursery herds. Mothers continue to nurse their calves for over a year, which may help to ensure that they devote at least two years' attention to each calf and do not become impregnated again as soon as the bulls return.

Although we know, from taggings and the painstaking monitoring of the comings and goings of different whales with the seasons, that whales can make long journeys through the oceans, we do not really have any idea of how they know where they are or where to go. Birds, like whales, migrate over huge distances, but with birds it is a relatively easy matter to probe their sense of direction. It has been shown that birds can use the sun and stars as a compass, and together with an internal biological clock they can also be used as a map. Birds also seem to have several back-up systems. They appear able to use the earth's magnetic field, low frequency sounds, ultra-violet radiation, the polarisation of the sky, the scent of the wind, and a whole host of other clues as to their whereabouts, although the heavenly compasses are their main instruments. These capabilities were demonstrated with a variety of techniques, many of which depend on moving the animals bodily to places where they would not ordinarily be, and then watching to see how they cope. Whales are hardly amenable to this approach. As Margaret Klinowska, a physiologist at Cambridge University, ruefully noted: "It is not practical to remove large numbers of Cetaceans from one area and release them in another."[9] Klinowska reckons that almost all we know about animal navigation comes from mistakes the animals make, either genuine mistakes or ones caused by interfering experimenters.

[9] Klinowska (1986) p 401.

The biggest mistake an aquatic animal can make is to run into land.

There is, fortunately, an excellent body of data at hand, the records kept by the British Museum (Natural History). Since 1324, when Edward II took them as a royal prerogative, whales and sturgeon have been "royal fish". Any that come ashore, either under their own misguided power or on the tides, are the property of the reigning monarch, and must be reported to the Receiver of Wrecks by whoever finds them. For the past 70 years or so the Natural History Museum in London has been collecting details of every whale stranded on British shores—a total of some 3,000 events. Most involve dead bodies, washed up by the currents. Interesting though these are, and in Victorian times they often attracted tourists and sightseers from miles around, they do not offer much insight into navigation. Some strandings, however, are of live animals, but not many, to be sure. Of more than 3,000 events, only 136 involved live animals. They may be in groups, so-called mass strandings. They may be single animals. The beached whales may die, or they may successfully be returned to the sea. Whatever the circumstances, live strandings are clearly mistakes.

Or are they? Stranded whales often behave quite perversely. After being guided back out to sea by well-meaning people they often turn right back and strand in the same spot, or just along the coast. This has led some people to claim that the whales are deliberately coming ashore to commit suicide. Other, perhaps less far-fetched ideas, include the suggestion that whales are tired, and enter shallow water to rest; or they are itchy, troubled by parasites, and come inshore to scratch; or they have reverted to some ancient mammalian memory that makes them seek safety on land; or sonar echoes get confused where the bottom is a gently sloping beach; or there are parasites in the whale's inner ear, making it deaf to its own sonar; or the whale is attempting to use an ancient migration route now closed by geological changes; or pollution; or radar, TV and radio emissions; or any one of a whole host of other factors that some people claim would cause whales to beach themselves.

Even a brief look at the Natural History Museum's records will discount many of these ideas. There does not seem to have been any increase in the number of live strandings over the past few decades, so it seems highly unlikely that any activity of modern man is to blame. As to problems with sonar, only two-thirds of the strandings in the UK take place on the gentle sloping shores likely to be confusing. The rest are on steep shores that ought to give good echoes. And although there may be parasites in the inner ear of stranded animals, not all stranded whales are infected. Furthermore, baleen whales, which do not use echolocation in the same way as toothed whales, nevertheless strand themselves. Cetaceans do not generally rest in shallow water, and although some groups of killer whales do seem to come to specific areas for a scratch, their behaviour never seems to result in stranding. Other groups of killer whales, and the rest of the Cetaceans, do not seem to scratch, and yet they do strand. Calling on suicide, or ancestral memories, does not get us anywhere. Perhaps strandings really are mistakes. What can the records tell us?

The most curious fact is that live strandings occur at discrete points around the British coast. Washed up bodies are found everywhere, and the records mirror such things as the number of coastguards nearby to spot a carcass. But live strandings are definitely concentrated at certain spots, independent of the people who might be present to witness them. Inshore species, such as the harbour porpoise, have rather few live strandings compared to the number of their dead bodies found, while species that normally dwell offshore strand relatively more often. Only 6 per cent of all harbour porpoise strandings are live strandings, while half the strandings of whitesided dolphins, an open-water whale, were live. This suggests that the offshore animals are outside their local area, relying less on their local knowledge and memory and more on their navigation systems, and hence more liable to make mistakes.

Margaret Klinowska thought that magnetism might perhaps be involved. We are used to thinking of the earth's magnetic field as a rather simple affair, with the magnetic North Pole somewhere near the true North Pole and a smooth field connecting the North and South Poles. In fact, though, it is much more complicated than that. There is a magnetic topography, just as there is a geological topography, with hills and dales, narrow valleys and broad plains. The magnetic topography is influenced by all sorts of features—a deposit of iron ore, for example, might make a magnetic mountain—and as far as the magnetic field is concerned the seashore is an invisible boundary.

Surveyors can plot the strength of the magnetic field and draw a magnetic contour map, which like a geological contour map shows the lie of the magnetic land. When Klinowska plotted live strandings on a magnetic contour map she found a clear link. "Live strandings occur *exclusively* where . . . valleys in the local geomagnetic field cross the coast or are blocked by islands."[10] Every single live stranding occurred at a place where magnetic contour lines crossed perpendicular to the coast line. Dead strandings occurred everywhere equally, influenced more by tidal flow than magnetic field.

Klinowska thinks that whales use the magnetic contours as a motorway system, staying in the magnetic valleys for the most part as they fin their way across the oceans of the world. "Live strandings, then, are equivalent to road accidents. The young, the old, the sick, the healthy, animals alone, animals in groups —accidents can happen to anybody."[11] If they are accidents, that would explain why the animals are often in shock, and need help to leave the beach. And it might explain why they often come to grief again at the next magnetic trap along the coast. "They are travelling in what seems to them an appropriate direction and the fact that there is another beach in the way comes as another nasty surprise."[12]

Klinowska's map-making certainly makes sense of the Cetacean strandings around the British coast—and wherever else she has looked—but it does not answer two further questions: where do the animals make their mistake, taking the wrong turn that leads them into unknown waters? And why does the accident happen in the first place?

The picture I have drawn, of a magnetic map that is essentially like an ordinary contour map, is incomplete. In addition to the contours, the topography determined by the earth and the magnet within it, there is an overlying ocean of magnetism caused by magnetic ions jetting around in the upper reaches of the atmosphere. This affects the magnetic field in two respects. Firstly, the overall strength of the magnetic field fluctuates with

[10] Klinowska (1986) p 413.
[11] Klinowska (1987) p 48.
[12] Klinowska (1987) p 48.

a daily rhythm. Magnetic ions in the atmosphere stream back and forth as the air heats up during the day and cools down at night. This causes a gentle swell in the height of the magnetic "sea level", although it does not affect the contours. Likewise the sun and moon have a tidal effect on the magnetic field, just as they do on the real oceans. Second, and more important to any animal trying to read the magnetic map, the regular harmonic fluctuation is occasionally upset by a pulse of magnetic radiation caused by an outburst of solar activity. These are violent storms on the generally calm magnetic sea, but they are more predictable than ordinary storms; dawn and the early morning are magnetically quiet times, while upheavals are more frequent after midday and in the early evening.

Magnetic contours offer whales a map in the featureless oceans. The regular shifts in the magnetic field, especially the early morning minimum, could provide them with information about the time of day and the season of the year. Certainly they do so for other animals, from honeybees to hamsters and humans. Together, the clock and map information in the earth's magnetic field would allow whales to find their way about. The topography is the road system, whales following prominent magnetic valleys just as we follow motorways. Of course they would have to learn the details of the magnetic routes they use, but calves could do that as they migrated with their mothers. Each morning the whale would reset its internal clock, and knowing roughly how fast it had been swimming would know how far it had come, and thus where it was. Magnetic storms, however, would temporarily destroy the system by interfering with the clock.

Because magnetic storms affect such things as radar and short-wave radio transmissions, military authorities are very interested in them. That means that there are excellent records of the strength of the magnetic field throughout the day, and storms are easy to isolate. Klinowska discovered that there was a clear link between whales stranding themselves and magnetic storms in the early morning, but the details depended on where the strandings took place. In the Irish Sea and the North Sea the whale tended to come ashore on the day of the disturbance. Along the south coast of England, the stranding came two days after the magnetic storm, while in Scotland and northern coasts the whales tended to strand about a day and a half after the storm. A look at the magnetic map revealed two major cross-roads, where two valleys intersect, out in the Atlantic. One is about two days' steady swimming time from the south coast. The other is about a day and a half from the Scottish coast. In the North and Irish Seas the animals are always within a day of land.

So it seems that whales use the earth's magnetic field as clock, compass and map. Almost all the time the system works very well, but occasionally there is a magnetic disturbance that upsets the whale's morning adjustment of its clock. If this happens on a day when the whale has to change course, it may cause the animal to make a mistake. It then continues on its erroneous way until it hits land. Klinowska's theory seems to solve the mystery. All we need to know now is how the whale detects not only the magnetic contours but also the relatively tiny fluctuations that signal a new day.

A magnetic sense would certainly help Cetaceans find their way as they made their long migratory journeys, but whales also need to find their food, and one another. Eyes work poorly underwater, for the particles in the water absorb the light. Sound, however, travels better underwater than in air, and many whales have an exquisite sense of hearing.

Sounds of human origin can confuse great whales. After the US invasion of Grenada, whale researchers reported that "military sonar signals apparently silenced and scattered the whales".[13] In the calving lagoons of Baja California, Gray whales often approach the idling outboards of skiffs. When the engine is switched off the whales are far less likely to come in close. And in the waters off Sri Lanka, researchers aboard the *Tulip*, a sailing yacht studying Sperm whales, reported that a newborn Sperm whale approached the region of the hull where the echo-sounder was emitting its steady train of pings.

Far more than sight, sound brings information about the world around to the whale. Sometimes the whale merely listens, and can probably detect fish and other foods that might make particular noises. The toothed whales, however, take a more active role; they use sound to echolocate. Like bats, they inhabit a dim and murky world in which light is not much use. Like

[13] Watkins, Moore and Tyack (1985) p 1.

bats, they use sounds to provide a picture of the world around them. Trains of high-frequency sounds—clicks, whistles, buzzes and the like—echo back from objects in the vicinity. The animal's brain then uses the echoes to compute the nature and position of objects around it.

It is easy to be overawed by this strange sense. What kind of ability is it that permits a blindfold animal to distinguish a fish it dislikes from one that it favours at a distance of 50 metres? In truth, though, I don't suppose the dolphin thinks it is all that marvellous. Dolphin scientists, were there any, might feel the same way about our "uncanny" ability to use invisible electro-magnetic radiation to form a picture of our environment. I suspect that the closest we can come to understanding dol-phins, and for that matter bats, is to assume that echolocation, for them, is rather like visual location for us.

Of the great whales, only Sperm whales definitely use sonar, to find their prey in the lightless depths. They feed by day and by night, but it makes no difference, for 300 metres below the surface all is inky black. The whales are generally in pursuit of squid. Sometimes they hover at great depth, apparently motionless, perhaps listening for their prey. More often they cruise the deeps, and as they search for their food they emit a series of high-powered clicks. It is the echoes of these clicks, returning from anything that might be in the way, that tell the whale whether there is food nearby. If there is, the clicks come more quickly, until the whale begins to sound like a creaking door. Then, silence. The hunt is over.

The clicks enable the whale to home in on its prey in the dark. They may even help with capture. Kenneth Norris, a widely respected authority on whales, and a professor at the Univer-sity of California at Santa Cruz, reckons that the clicks may be powerful enough to stun the prey, making it easier to grab.

Sperm whales make other sounds too. Most intriguing of these are the so-called codas. These are distinctive patterns of clicks that seem to be different for different individuals. They are heard most often when the whales are lolling at the surface socialising, and it is hard not to imagine that they represent some form of communication between the animals. In addition, one particular type of coda, a pattern of five clicks, would appear to have a distinct meaning. It seems to be used by whales as they join or leave a group, and might be a kind of warning, a signal of status.

Humpbacks may or may not be the noisiest of the great whales, but they are certainly the most studied. In all likelihood that is the result of their uniquely evocative song. Few people have heard the song of the Humpback for real, in the crystal waters of the Bahamas, or in the Hawaiian chain, or off the Cape Verde islands off west Africa. When you do hear it, issuing mysteriously from the very walls of the vessel you are in, there is something magical about the sound. But that is not an experience granted to many. Nevertheless, thanks to some astute promotion, the song of the Humpback, with its eerie plaintive notes, has become one of the instantly recognised noises of nature.

Musicians were quick to sense the melodious possibilities of whale song, and people as disparate as Judy Collins and Pink Floyd co-opted Humpback singers into their recordings. Jazz musicians, such as Paul Winter, improvised along with whales. I had to smother a laugh during a screening of the film *Legend* when I realised that the "neighing" of a pair of friendly unicorns was in fact the cry of a Humpback whale. And in 1970, whales became the stars of the largest single pressing in the history of the record industry. Capitol Records manufactured 10 million copies of a record, which *National Geographic* magazine gave away to each of its subscribers, who listened astonished to the songs of the Humpback whale.

Humpback song may have attracted the most attention, but it is just one of the noises these mighty mammals make. In general Cetaceans use sound for two purposes: communications and intelligence. Song broadcasts a message about the singer. What of the Humpback's other noises?

Scientists have explored the echolocation of the smaller toothed Cetaceans to the very limits of their capabilities, largely because dolphins and porpoises can be kept in captivity relatively easily. That enables scientists to control the experiments, to make sure that sound, and sound alone, provides information.

For example, only an experiment could prove that dolphins hear through their jaws. Their ears are tiny, mere pinpricks behind the eye, but there are several things about the lower jaw that suggested it was the route for echoes returning from the surroundings. The jaw is hollow, unusual among mammals, and contains fatty molecules that would probably help conduct sound to the ear. When physiologists stimulated the jaw, they got a response in the hearing centre of the brain. But all that is circumstantial. Until Randy Brill, head trainer at Brookfield Zoo in Chicago, taught his star performer to wear a foam rubber "hood" over his lower jaw, there was no proof. By careful and sympathetic training, Brill persuaded Nemo, a 13-year-old Bottlenose dolphin, to wear the hood. Without it, Nemo could easily tell the difference between two targets, even when securely blindfolded. When the hood was in place, cutting sound transmission through the jaw, Nemo could not use his echolocation system.

Experiments, like the ones on Nemo, are the only way to be certain, and although recordings of baleen whales, including Humpbacks, have revealed that they do occasionally make noises that could perhaps be used for echolocation, there have not been the same opportunities to find out for certain. In July 1976, however, there was a chance to study a full grown Humpback whale.

Adult Humpbacks occasionally get themselves entangled in fishermen's nets off the coast of Newfoundland. Usually these whales drown or are killed by the fishermen, but in this case somebody called in Peter Beamish, a scientist working at the Bedford Institute in Nova Scotia. Beamish freed the whale and fitted it with a padded harness. He then surrounded it with a relatively large net cage, and the whale swam about in its enclosure with no apparent discomfort. Beamish arranged a set of aluminium poles in the enclosure, and over the next few days tested the Humpback to see whether it could avoid the poles using echolocation.

The first tests were done during the daytime, and the whale had no trouble in avoiding the poles. Beamish was listening in on a hydrophone, and heard nothing. As the whale was silent, Beamish concluded that it was using its eyes to avoid the obstacles. Next Beamish slipped a foam-lined blindfold over the whale's eyes. Still it remained silent, but this time it crashed into the poles. Other tests were conducted, but at no time did the whale make any noises while navigating the maze. If there was light, and if it could see, it avoided the poles. If it was dark, or if it could not see, the whale blundered. Eventually the whale was released unharmed, having failed to persuade Beamish that it was capable of echolocation. That is still the state of play with the baleen whales; there is no conclusive evidence that they use echolocation to find their way about, though there are strong suggestions that they can use low-frequency sounds to identify larger objects, such as ships or other whales, or to identify features on the sea bottom. Definitive research is still needed to discover whether Humpbacks can echolocate.

Echolocation, however, was never the Humpback's main claim to fame. Its song is a different matter. Whalers had known about the Humpback's song for some time, but made little of it in their logs. It was not until World War II that people began listening to the underwater world in earnest, and then whale song was a distraction. Indeed, in the early days of the US Navy's Sound Fixing and Ranging (SOFAR) stations nobody knew what they were hearing. SOFAR stations were listening for enemy submarines, and submarines too were equipped with underwater ears, but some of the strange underwater sounds baffled everybody, including the zoologists who had never previously thought to listen in to sounds underwater. In any case, the submarines were themselves so noisy that the signals from the early hydrophones had to be electronically filtered to get rid of the listening vessel's own engine rumblings. These filters cut out many of the low-frequency noises made by whales. After the war a new hydrophone, the BQR, was developed, so sensitive that submarines had perforce to be silenced. Filters were now unnecessary, and whale sounds came flooding into the listeners' ears.

By 1952 O. W. Schreiber reported in the *Journal of the Acoustical Society of America* that the SOFAR station in Hawaii had recorded sounds with "a rather musical quality", and that these sounds were produced only when there were whales in the area, which fact "has led to the belief that they are produced by whales".[14] Seventeen years later the whale biologist Roger Payne extended this finding. Payne, working off the Bermudas, discovered that the Humpback's song was a long sequence with a formal pattern, and that it was repeated over and over again. It was, quite clearly, a song in exactly the same sense that a blackbird's notes are a song.

[14] Quoted by Winn and Winn (1985) p 94.

50

Lois and Howard Winn, whale biologists at the University of Rhode Island, watched and recorded Humpbacks singing around Puerto Rico and the Virgin Islands. They had a directional hydrophone, which allowed them to take bearings on a singing Humpback and sneak up on it in their research vessel, *Trident*. They discovered that songs lasted between six and twenty minutes, and one occasion a whale repeated his song for at least twenty-four hours. Singers were always alone, and lone whales were almost always singing. The Winns thought the singers might be males, singing to announce their presence to females and to keep rivals away, but no one had ever tried to sex Humpbacks at sea. Living cells from the singer would give the answer, for the sex chromosomes are easily distinguished, and to get a sample of skin cells the Winns used a dart gun originally developed for pinning down the Loch Ness monster. The gun fired a hollow needle, about six inches long and a quarter of an inch across. The needle was thoughtfully covered with antiseptic salve to prevent infection, and removed a plug of skin from the Humpback's hide. With this device, the Winns went in pursuit of singing Humpbacks, but the odds were against them. Singers are very wary, as the old whalers knew, and the weather was bad too. After an extended cruise they had obtained only two skin samples from singing Humpbacks. But both of them were males.

Conclusive proof that all Humpback singers are males had to wait another few years, and involved a much more direct approach. Flip Nicklin, an underwater photographer with a team of Roger Payne's in the Hawaiian islands, trained himself to hold his breath as he scuba'd down to a singing whale at perhaps 40 metres. Singers hang motionless 15 to 30 metres below the surface, head down and tail up, at a 45 degree angle. Nicklin approached the singer from the rear and shot off as many frames as he could of the tail and genital area before giving vent to the stream of bubbles that might otherwise have disturbed the whale prematurely. Every one of the whales was male, and the team also made a series of observations that helped to explain why Humpbacks sing.

The song itself is a curious thing. A singer will repeat the same sequence of notes—variously characterised as moans, cries, chirps, yups, ooos, ratchets and snores—over and over again. The songs differ slightly between individuals, and much

51

more so between areas, so that the sound of a Hawaiian Humpback is recognisably different from that of one from Baja California, and Pacific animals sing differently from Atlantic ones. Katy Payne discovered that each year the songs change to some degree. The changes are not marked, a syllable here, a repetition there, and they occur during the singing season, not between seasons. All whales seem to track the changes, altering their songs as the season progresses, but why they do so is a mystery. The Winns suggest that it may be that a new song each year is more stimulating to the females, or that a dominant male calls the tune and the others follow his lead.

Whatever the explanation, the question of why Humpbacks sing remains. One possibility is easily dealt with. It has been suggested that, with their complicated songs and vast brains, Humpbacks are talking to one another, using a language in the same way that people do. "This," as Lois and Howard Winn say, "is surely not so. Although the song contains considerable information, everything suggests that it is information of a rather simple kind ... Birds sing in much the same way, varying the order and linkage of a similar number of syllables into various song types. But the songs of some birds have many more syllables than the Humpback whale's song, and their pattern is not so fixed."[15]

Why, then, do they sing? Roger Payne says that "singing is probably an aspect of courtship"[16] and this is almost certainly correct.

Jim Darling, a zoologist with the West Coast Whale Research Foundation in Vancouver, was part of Payne's research team in Hawaii, and has put in a great deal of effort to understand the song of the Humpback. As Flip Nicklin had proved, singing Humpbacks were invariably male. A singer might sing for hours until one of two things happened: either another adult, not a singer, joined the singer, or he stopped and rushed off to join a larger group of males pursuing a female in oestrus.

Male birds often use song to attract females to them. Darling thought that Humpbacks might be doing the same, but poring over reams of identification photographs he discovered that

[15] Winn and Winn (1985) pp 97–8.
[16] Payne (1982) p 466.

every one of the whales who approached a singer was also a male. This was one element of the puzzle.

Another was that Humpback males fight violently with one another but this does not happen often. Usually, when one male joins another who is singing, they will simply swim apart. They may slap flippers together, and very rarely they will go in for stronger blows, usually from the tail. Males escorting a female also fight quite violently. Darling admits that it took a long time—many seasons of observation—to discover this. "I do not know why it took us so long," he admits. "Partly because we did not like to think of whales as fighting."[17] But fight they certainly did, often inflicting bloody wounds with violent tail lashings. The males seemed to exchange threats too. They blew long streams of bubbles from their blowholes, or arched the back and stuck the head out of the water. While in this position the whale sometimes gulped air, distending the throat pleats and perhaps, Darling speculates, making itself look bigger to rivals. That put the song in a different light, especially as it attracted other males, not females.

Song, Darling thinks, is the Humpback's equivalent of the red deer's antlers, although unlike antlers, song is not used as a weapon. It is purely a display that establishes an animal's position in the dominance hierarchy. Generally, disputes are easily settled, as each whale can size up his opponent by means of their songs. Only if two animals are closely matched will they go on to a physical fight. The mysterious changes in songs during the season and from year to year ought perhaps to be regarded as akin to growth, more dominant animals leading the way with more complex songs.

Whether Darling is correct or not, and at this stage it is hard to be sure either way, his hypothesis does at least account for the facts of Humpback behaviour. It has taken a lot of research to get this far. It will take a lot more to show, for example, that younger or smaller males lag behind in the growth of their song, one clear prediction of the hypothesis. Many other unknowns remain, among them one that seems to have been almost completely overlooked in the rush to understand the song: Humpbacks have no vocal cords, so how do they produce their various sounds, including those that make up the songs? "The

17 Darling (1984) p 8.

53

best guess is that they are produced by the various valves, sacs, and muscles found in the whale's larynx and in its respiratory tract."[18] Indeed, but that doesn't add much to what we already know.

Intimate knowledge of Humpback songs has thrown light on another mystery—that of migration. Roger Payne is fond of telling how he sat one day listening to a tape of singing whales made by friends in Baja California. He was completely familiar with the songs of Hawaiian whales, and was amazed to recognise notes, cadences and passages that he thought were peculiar to the Hawaiian whales. He checked, but there was no mistake; the whales had been recorded across the Pacific, off Mexico. There could be only one explanation: the whales taped in Baja California had learned their songs in Hawaii. But that was not possible, for scientists had long been certain that whales who wintered and calved in different areas followed different migration routes to different feeding grounds. The whales in Hawaii were supposed to spend their summers around the Aleutian Islands, while those from Baja California were supposed to feed in more easterly areas, perhaps off southern Alaska. The two were not supposed to mingle. So how was a Humpback singing Hawaiian songs off Mexico?

Mug shots—or rather tail shots—proved that the scientists had been wrong. Jim Darling had compiled his collection of photographs of the tail flukes of Humpbacks in Hawaii. The scars, markings and outline of a Humpback's tail are as distinctive as our own faces. Darling could identify about 350 whales. Charles and Virginia Jurasz had similar files of the whales that gathered off southern Alaska. When Jim and the Juraszes compared their photographs they found not one but seven whales that definitely wintered off Hawaii and summered off southern Alaska. So at least some of the mid-ocean whales did not go to the Aleutians.

Katy Payne, then Roger's wife, went to the Revillagigedo Islands in Baja California to photograph the Humpbacks there, which had caused the upset in the first place. They got good shots of eleven animals, which Jim Darling then compared with those in his files. Two of the eleven had been positively identified on the other side of the Pacific, off the island of Maui. The

[18] Winn and Winn (1985) p 97.

case was closed. Humpbacks do not stay in neat little groups, shuttling back and forth between set summer and winter grounds. They switch between stocks, and can range far more widely than was previously thought.

This is a good example of the way in which meticulous and thorough work by whale scientists around the world has filled in another small detail of our picture of whale life. There is much still to be discovered, but one question remains. How smart are whales?

In the campaign to save the whales, much is made of their huge brains and, naturally, huge intelligence. Whales are the only animals, apart from elephants, to have brains larger than our own. The great whales have great brains, but that is only to be expected given how large the animals themselves are. Size, however, is not nearly as important as complexity and it is interesting that among the Cetaceans the Sperm whales and orcas have the most complexly convoluted brains. As predators, and with a well-defined social system, Sperm whales and orcas would need a lot of brain-power, and there is a suggestion that their brains may even be more complex than our own. But the baleen whales are almost certainly not as clever as they have been painted. Whalemen say that Finback whales can be hunted with the same sorts of techniques one might use to go after a herd of bison or caribou. Groups can be stampeded, and during the chase the captain can pick off individual whales by tricking them into diving and turning in specific directions. The whaler takes advantage of these predictable reactions in order to put the ship near the whale when it surfaces. This does not speak of enormous intelligence, and nor would we expect baleen whales, knowing what we do of their way of life, to have that much need for the kind of flexible thought processes that we value so highly. The whole question of whale intelligence is still rather doubtful.

What is not in doubt is that whales are the largest animals ever to have lived on this planet.[19] Superlatives attract further

[19] Recent reports (*New Scientist*, 23 April 1987, p 24) from the deserts of New Mexico suggest that a brontosaurus-like dinosaur—called Seismosaurus—might have been longer than the Blue whale, an estimated 40 metres from head to tail. But it would have weighed only 40 tonnes, far less than a Blue whale, and so far only eight tail bones of the earth shaker have been found.

exaggeration, and it can be hard to separate facts from fantasies, documented measurements taken in the approved manner from fishy tales. The records show that the largest Blue whale stretched 29.87 metres from the tip of her snout to the notch between her tail flukes. Dr Masaharu Nishiwaki measured her carcass at the Japanese Whale Research Institute in the Antarctic during the 1946/7 season. The heaviest was a 27.12 metres female cut up on a Japanese floating factory on 27 January 1948. Her butchered carcass weighed in at 129.5 tonnes, which allowing for a 12 per cent loss of blood and body fluids meant that she probably weighed 145 tonnes alive.

Even much smaller whales are still a lot of meat. After more than 60 million years of evolution Cetaceans in general, and the great whales in particular, are supremely well adapted to making their living in the water. With the exception of the Sperm whale they are animated suction devices, skimming vast numbers of tiny animals from the sea and turning them into whale meat. As such they represented an irresistible target for people in search of protein.

History

They sought it with thimbles, they sought it with care;
    They pursued it with forks and hope;
They threatened its life with a railway-share;
    They charmed it with smiles and soap.

                    Fit the Fifth: *The Beaver's Lesson*

People living on the coast have probably exploited whales since
before history. Dead whales would have been washed ashore,
and occasionally people would be confronted with the mysterious phenomenon of whale strandings, in which for reasons
unfathomable the animals beached themselves and died.
Blocks of blubber cut from these lucky finds provided fuel for
heat and light and could be buried for future needs. The meat
might have been eaten if the whale was not too long dead. Given
the richness of the harvest, it is not surprising that some people
set out actively to hunt whales from small boats, still the way of
life of Eskimos in the high Arctic.

This hunting, however, was not for material gain. It was
strictly subsistence hunting, satisfying the needs of local people.
Not just the meat and blubber but all parts of the whale found a
use among these hunters.

The first records we have of a more commercial approach to
whaling come from the Basque country between France and
Spain and date to about 900 AD. On a sailor's chart you will see
that a tongue of deep water licks close into the southeast corner
of the Bay of Biscay, right at the foot of the Pyrénées. Whales
generally prefer deep water, but that trench brought them
within sight of the people on the coast. The Basque whalers put
out to sea in small boats in pursuit of the Biscayan Right
whales. They were the right whales to pursue because they
swam slowly and floated when dead.

Lookouts kept a close watch on the water, and when whales
were sighted the whole village co-operated in the hunt, some
going out to drive the whales closer inshore, others going in for
the kill. These first whalers used hand-held harpoons to spear

their prey, and because their boats were so frail the Basques attached the harpoon line to floating drogues. The whalers followed the whale until it was exhausted and they could kill it. Then they rowed their prize back to shore where the blubber was flensed from the carcass and rendered in giant cauldrons. Scraps of blubber fed the fires below the blackened copper pots, boiling out the oil.

This whale oil was a valuable trade commodity, and each villager got a share of the proceeds, according to the number of harpoons landed in the whale. Whale oil was used mostly in lamps, although later it became the mainstay of the soap industry; the shift in fashion from the simple ruff of the Elizabethans to the much more complex lace collars of later years probably owes as much to cheaper whale oil—in soap and lacemakers' lamps—as to anything else.

Whale meat was not eaten much. The tongue was considered a delicacy, and in the sixteenth century was sold, along with some of the other meat, in local markets around Biarritz.

Records of whaling at this early period are sketchy at best. Certainly by the thirteenth century the trade flourished off the town of Biarritz. The town's seal shows a whale being harpooned. Before that, in 1197, King John, in his capacity as Duke of Guyenne, made a canny trade with a nobleman, Vital de Biole. King John gave Vital de Biole fifty angevin livres, to be levied on the first two whales taken at Biarritz each season. In exchange, Vital de Biole gave King John the right to fish off the island of Guernsey. As the whales had been free of all taxes before then, John got the Guernsey fishery essentially for nothing.

The whalers took their toll of Biscayan Right whales, which have never recovered, and to maintain their livelihood they were forced further afield. They ventured to Iceland, Spitsbergen and Greenland, all without benefit of compass, and by 1538 were in Newfoundland. The most important whale fishery, once the Bay of Biscay had been exhausted, was the so-called Greenland fishery, actually off the coast of Spitsbergen.

The Spitsbergen fishery was the scene of great battles between the English and the Dutch, who fought repeatedly over just who had rights over the island and its waters. The English based their claim on an expedition conducted by Sir Hugh Willoughby, who discovered Spitsbergen in 1553 while trying to find a northeast route to Cathay. Willoughby died in Lapland but his pilot-major, Richard Chancellor, continued the voyage and discovered "the kingdome of Moscovia by the North-east". The English established the Muscovy Company, the first of the great joint-stock companies formed for trade, to exploit these new lands. After another voyage of discovery in 1557 the Company received reports of "many whales, very monstrous, about our ships, some by estimation of sixty feet long, and being the ingendring time they roared and cried terriblie".[1]

The first Muscovy Company expedition reached Spitsbergen in 1610, and the captain was censured for bringing home blubber instead of the more valuable oil. Nevertheless, the dividend that year was 20 per cent. Thomas Edge, a captain with the Muscovy Company, provided one of the best descriptions of the whale and how whalers dealt with it at the time.

> The whale is a fish or sea-beast of a huge bignesse, about sixtie five foot long, and thirtie five foot thicke, his head is a third part of all his bodies quantity, his spacious mouth contayning a very great tongue, and all his finnes, which we call whale finnes. [These are the plates of whalebone.] These finnes are rooted in his upper chap, and spread over his tongue on both sides his mouth, being in number about two hundred and fiftie on one side, and as many on the other side. The longest finnes are placed in the midst of his mouth, and the rest doe shorten by their proportionable degrees, backward and forwards, from ten or eleven foot long to foure inches in length, his eyes are not much bigger than an Oxes eyes, his body is in fashion almost round forwards, growing on still narrower towards his tayle from his bellie; his tayle is about twentie foot broad, and of a tough solide substance, which we use for chockes to chop the blubber on (which yields oyle), and of like nature are his two swimming finnes (and they grow forward on him). This creature commeth oftentimes above the water, spouting eight or nine times

[1] Jenkins (1921) p 71.

before he goeth down againe, whereby he may be descried two or three leagues off.[2]

To his description of the whale, Edge added an account of the hunt, beautifully illustrated with engravings. After striking the whale with the harpoon the whalers attempt to lance him:

> [In] lancing him they strike neere the finnes he swimmeth withall, and as lowe under water neere his bellie as conveniently they can; but when he is lanced he friskes and strikes with his tayle so forcibly, that many times when he hitteth a shallop he splitteth her in pieces.
>
> The whale having received his deadly wound, then he spouteth blood (whereas formerly he cast forth water) and his strength beginneth to fayle him.

This gives the men a chance to tow the whale back to the ship, where it is laid across the stern for the blubber to be cut off.

> [T]here is a crane or capstan placed purposely upon the poope of the ship, from whence there descendeth a rope with a hook in it; this hooke is made to take hold on a piece of blubber; and as the men wind the capstan, so the cutter with his long knife looseth the fat from the flesh, even as if the lard of a swine were to be cut off from the leane.[3]

The blubber was then rowed ashore to the cookeries, where it was boiled to extract the oil. The fins, the whalebone plates, were severed one from another with axes, cleaned and bundled up into packages of fifty.

The Dutch, meanwhile, simply ignored the English claim on Spitsbergen and went ahead and exploited the whales there. In 1612, two years after the first Muscovy Company expedition, the English whalers found Dutch ships alongside them in the ice. They had been guided there by Allan Sallowes, an Englishman who had worked for the Muscovy Company for twenty years. He had to flee England in debt and ran to Holland, where the Dutch employed him to organise their own whaling industry. This he did, bringing in harpooners from the Bay of Biscay.

In 1614, it looked as if there might be trouble between English

[2] Jenkins (1921) pp 116–7.
[3] Jenkins (1921) p 117.

and Dutch in Spitsbergen. The Dutch formed a big fleet of fourteen whalers, levying a tax on each of the whalers to pay for four men of war to defend them. The English too set forth in strength, with fifteen ships, all equipped with artillery as well as fishing gear. In the event, nothing happened and the two fleets fished, rather unsuccessfully, without fighting. Three years later, in 1617, Thomas Edge had become commander of the Muscovy Company's whaling fleet. He heard that the Dutch were whaling a little way away, in Horn Sound, and sent his Vice-Admiral there to "put the Flemmings from thence and take what they had gotten". This they did, much to Edge's dissatisfaction, since the goods taken were worth less than £20. Another English ship, under Captain William Heley, was luckier. They too attacked a Dutch whaler in Horn Sound and confiscated "two hundred hogsheads of blubber and two whales and a half to cut up, a great copper, and divers other provisions, and sent him away ballasted with stones."[4] The Dutch were not amused.

The following year the Dutch sent twenty-three ships to Spitsbergen. They had two boats to each English one, and fished alongside the English "with a full purpose to drive the English from their Harbours, and to revenge the injurie done them the yeere before". On 24 June 1618 Master Robert Salmon wrote home about the set-to. He tells how they had killed thirteen whales, but got little oil as a result of difficulties working in the ice, and then goes on:

> Here is five sayle of Flemmings which have fourteen and sixteene pieces of Ordnance in a ship; and they doe man out eighteen shallops so that with theirs and ours there is thirtie shallops in the bay, too many for us to make voyage; there is at least fifteene hundred tunnes of shipping of the Flemmings; we have reasonable good quarter with them, for we are merry aboord of them, and they of us, they have a good store of Sacks, and are very kinde to us.

But he was also a realist, for he adds:

> The Company must take another course the next yeere if they mean to make any benefit of this country, they must send better ships that must beat these knaves out of this country.[5]

[4] Jenkins (1921) p 106.
[5] Jenkins (1921) p 106.

Five days later (29 June) Master Sherwin also wrote home about the Dutch. "Let them all go hang themselves, and although you be not strong enough to meddle with them, yet the worst words are too good for them, the time may come you may be revenged upon them againe." Sherwin was approached by two Dutch ships, but handled them carefully "for fear of after-claps". He said that had it been later in the year (with the proceeds gained?) "we would have handled them better. Now they be gone for Horne Sound, I would that they had all of them as good a pair of hornes growing on their heads, as is in this country."

William Heley, who had been so lucky the year before when he confiscated the two hundred hogsheads of blubber, found himself in big trouble. On 19 July, five Dutch ships, well armed with "cast pieces with brass bases and murtherers" set upon Heley in the *Pleasure*, which was in company with two other English ships. Despite offering resistance the English were forced to anchor or run ashore, where their ships were rifled and their casks burned. The English fleet then dispersed, their voyage being "utterly overthrowne" and returned empty. The Muscovy Company put the loss at £66,000, besides ships and men.

The English and the Dutch continued their squabbles in the courts and through diplomatic channels, but no good really came of it. The simple fact is that the Dutch continued to do very well out of the Spitsbergen fishery while the English failed again and again. Whales became scarcer in the bays and the Dutch Noordsche Company sent out voyages of discovery, looking for more whales. They found them in the ice to the west of Spitsbergen, and began to exploit that stock, but the main effort remained in East Spitsbergen, centred on a shore station called Blubbertown. There were eight or ten oil-coppers, warehouses, and huts for the thousand or more whalers who came up each year. Merchants found it worthwhile voyaging north, selling mostly brandy and tobacco, and a baker set up shop. Each morning he sounded a horn when the hot rolls and bread were ready. In 1633 Blubbertown was thriving. By 1639 it was already falling into disrepair. The reason? As in the Bay of Biscay, the whalers had taken too many whales. The stock was exhausted.

While the Dutch continued to profit from the northern whale

fishery, the English were extremely unsuccessful. In the ten
years from 1699 to 1708 the Dutch sent out 1,652 ships, which
caught 8,537 whales. These sold for the equivalent of almost
£2.5 million of which £425,000—worth almost £15 million in
1987—was clear gain. The English had almost nothing. They
formed a new Greenland Company in 1696, which proceeded to
lose money at a phenomenal rate. Six reasons were given in a
document issued in 1722, when more money for whaling was
being sought.

1) Ships were commanded by persons unacquainted with
the business, who interfered with the fishery, whereas the
chief harpooner ought to have commanded at this time.
2) Captains had fixed pay, but should have been on a share
of the whales taken.
3) Blubber taken home was slovenly and wastefully managed in boiling, and the fins were ill cleaned, so that the
products offered for sale only fetched an inferior price.
4) The lines and fishing instruments were injured from want
of care and frequently embezzled.
5) The ships were extravagantly fitted; an exorbitant price
paid for materials and large sums spent on incidentals,
which ought to have been saved.
6) The last ship sent out was unfortunately wrecked, after
securing eleven whales, a misfortune which accelerated the
ruin of the company.[6]

The Dutch had invented the wooden slipway, for drawing the
whales on shore, which made it easier to remove the blubber,
but that was easily copied and hardly enough to account for the
immense contrast between the two nations. In fact, the difference between the Dutch and the English seems mainly to have
been one of attitude. The Dutch (and some of the English
"interlopers") believed in free trade. The Muscovy Company,
and other companies that came afterwards, were monopolists.
The monopolists believed that, by virtue of their investment in
the voyages of discovery, and the exploitation of the resources
discovered, they should enjoy sole rights to those resources.
They wanted freedom from taxes too, and many other con-
cessions. The free-traders simply went about their business,

[6] Jenkins (1921) pp 163–4.

trying to make as much profit as they could. The arguments between the two factions, each explaining why its method of business was the only sensible one, have a curiously modern ring to them. And in the seventeenth and eighteenth centuries, the free-traders prospered while the monopolists went bankrupt, repeatedly.

The South Sea Company, established in 1711 to reduce the National Debt, decided to go whaling in 1724. At the same time George I granted a fresh monopoly on soap making to the Society of Soapmakers in the City of Westminster in the County of Middlesex. The Society had the power to search offenders, who might be making soap illicitly. They took their monopoly seriously, and issued proceedings against the old soap-boilers, who used fish oil and refused to have their soap tested and marked by the assay-master. "[A]nd who also, though not a body corporate, presumed to assemble in taverns in London and to confer about the sale of their soap and the buying of fish oil from the Greenland Company."

The South Sea Company could find no Englishmen skilled in whaling, and was forced to employ foreign labour; 152 skilled whalers from Holstein cost the Company £3,356, while 353 home-grown whalers were paid only £3,151, just one reason why the entire endeavour was a disaster. Calculations showed that each ship needed roughly three whales a year to be successful; the South Sea Company's ships averaged less than one whale a year. In eight years whaling the Company spent £262,172 9s 6d, and recovered £84,390 6s 6d from the sale of oil, ships and the like, for a total loss of £177,782 3s 0d.[7] (Fortunately for us they were better book-keepers than whalers.)

The English finally adopted a sort of free trade, encouraged by a government subsidy on each whaler, which made it worth-while to undertake the longer and more dangerous voyages needed now that the Spitsbergen stock was exhausted. The whale fleets of the North Sea ports—Hull, Whitby, and in Scotland Dundee and Peterhead—began to grow.

The surgeon of the Whitby ship *Volunteer* published an account of his voyage. The ship was of 400 tonnes and carried eight boats. The ship's company was sixty-three, and the

[7] The pound of that time was worth about 35 times the pound of 1987.

66

surgeon was paid £3 10s 0d per month, with an additional £1 1s 0d per fish. They left Whitby on 24 March, 1772 and on 26 April, up around 70 degrees N, saw two whales, one close to the ship. They were large, but not black like the Right whales:

> These kind of whales have fins on their backs, and are seldom if ever caught, it being dangerous to attempt it for as soon as they are struck they are so strong and swift in nature that no boats can get up to the assistance of the boat that is made fast to them before they are gone, and theire is great danger of the boats oversetting.

So there were still whales in the waters off Spitsbergen, but the whalers of the time could not exploit them.

> I never heard of any that attempted striking of that kind but a Dutchman some years since, but he was never more heard of, so that it was suspected the whale had run him off, and he had perished in the attempt.

The *Volunteer* returned to Whitby on 19 August, with five whales. The whales yielded 186 butts of blubber, which boiled down to 65 tonnes of oil. The oil sold at £20 a tonne, giving a total of £1,300. There were between four and five tonnes of whalebone too, providing a further £2,300. With the bounty money the total proceeds of the voyage amounted to some £4,800, just about profitable.

The Dutch and the Germans were still doing considerably better out of the Arctic fishery than the English. There is an excellent account from one F. G. Köhler, a sailmaker from Pirna who shipped on an Arctic voyage in 1801. Köhler did not publish his account until 1820, and he warns his readers on no account to take part in the Greenland fishery.

Köhler sailed on the *Greenland*, a three-masted vessel, in the company of seventeen other ships. They left from Altona on 16 March, 1801, and although the expedition was a German one, only five of the forty-two crew of the *Greenland* were Germans. The rest were "Dutch, Danes, and Jutlanders", and this is representative of German whaling of the time; mostly, it was conducted by foreigners.

The crew was divided into three watches, four hours on and eight hours off, and Köhler spares few details of life aboard. On the subject of food, he says:

There is no fear of making my readers' mouths water. At four o'clock in the morning we get coarse groats with some butter, and so one morning like another. Dinner shows very little variation. On Sunday grey peas and Stockfish, Tuesday grey peas and meat, Wednesday yellow peas and Stockfish, Thursday the same, Friday grey and meat, Saturday yellow and Stockfish; and so the loathsome grey and yellow change about one week with the other.

The crew rejoiced for several days when they got anything but peas; twice they got white beans and twice sauerkraut. On 28 May they had a feast, for it was the captain's birthday. There were twenty-two bottles of wine, and the crew drank the King of Denmark's health. The captain also supplied a few potatoes for some of the crew, and Köhler records how, for once, he was in luck, getting one whole potato and a piece of another. The bread was bad, and full of worms. It looked, says Köhler, like peat, and had to be washed before it could be eaten. The water was no better, since the water casks were filled with whale oil when empty, and after a very casual cleaning were used again for water the following year.

Bad though conditions were, whalers probably were no worse off than other sailors of the time. Their ships were overcrowded, poorly ventilated, and very wet in any kind of rough weather. The men usually had no change of clothing, and most suffered scurvy and other skin diseases, but if anything they were slightly better off than other seamen. They could at least change their diet occasionally.

Some did try whale meat, but preferred beef. Trapped crews, who were compelled to eat whale meat, took to it well enough. Whalers also tried seagulls, ducks, and enormous quantities of eggs. They sometimes shot reindeer, and even bear. Beer was the main drink, but each man also brought along his own supply of tea and coffee.

Journeys were enlivened by meetings with other boats, and Köhler describes the custom whereby the captains of ships that passed in the ice exchanged information with one another:

On these occasions I have often remarked the pride of the English. Every English ship waits until the other ship has first given its account of the fishing, so that they (the English, that is) always give a pair of fish in excess. On one occasion,

68

as I stood on the poop to give the signal our captain said, "Give the number ten and you will see that the English ship will announce eleven or twelve." And so it happened.

Records of the time show that the English had improved their abilities, and were doing quite well, though perhaps not always as well as they would like others to think. Köhler admired the English. "As seamen they are skilful navigators, and I have often observed with pleasure how on their ships they set to work with skill and agility."

Finally the *Greenland* sighted its first whale, off Spitsbergen. It now becomes obvious that Köhler is not exactly the heroic type of whaler. He says that he was so unfortunate as to be in the boat that set out to harpoon it, describing how his heart beat with foreboding, and how when they eventually came close enough to hear its blow his fear grew. As they approached, the whale became restive, stirred itself, and in the ensuing commotion made off. The captain was greatly disappointed, for he thought the whale to be worth about 8,000 thalers, but Köhler openly rejoiced: "My heart was joyful," he says.

Köhler's account is, I think, all the more interesting precisely because he is not all that keen on whaling. On one occasion a whale smashed three boats in its struggles. Köhler was in one of the others, and fought for twelve or sixteen hours before they killed it. During the struggle, Köhler complains, they had neither bread nor water, and thought every minute would be their last. That whale came at the end of the voyage, and provided sixty-four barrels of oil. The *Greenland* caught two others, one an easier capture that provided forty-five barrels, and the third a suckling that yielded just two barrels. They found a dead whale, and proceeded to flense it, at which Köhler complained of the stink. A crewman reassured him that the smell was quite bearable, and nothing compared to the stench of a dead whale they had come across on the previous voyage. That whale was so malodorous, said the old hand, that it caused the men's heads to swell up, a fishy tale but one that Köhler was inclined to believe. He really did not like the smell of whale, especially whale that had been dead some days, and spends many words on the topic.

The *Greenland*, with its small cargo, set sail for home on 23 August. Köhler says it was impossible to describe their feelings of joy. The ship's doctor broke into poetry to commemorate their farewell to the world of ice. They got to Heligoland, where they declined a pilot because his services were too expensive, and were chased by an English convoy. Fleeing from the English the *Greenland* ran aground, and was only floated off with some difficulty, but eventually got home safely with her cargo.

Köhler's pay for the entire voyage amounted to ten shillings. Perhaps not surprisingly, he went back to making sails.

By the 1820s the Greenland fishery was beginning to decline, and quite rapidly. This was the result of two factors. One was a change in the source of lighting in the towns of England. Coal gas had begun to overhaul gas made from whale oil. This was not because of the natural superiority of coal gas, at least not to a whaler such as G. W. Manby, who sailed from Liverpool in 1821. In his journal Manby comments on the new-fangled coal gas:

> The advantage of gas produced from oil, compared with that obtained from coal, is so great that it is astonishing that oil gas is not in general use. The gas from oil has no bad or disagreeable quality, it gives a far more brilliant light than the other, one cubic foot of gas from oil going as far as twice that quantity of coal gas, and it is, moreover, much cheaper. That from coal, on the contrary, is extremely offensive to the smell, dangerous to the health on being inhaled, and injurious to furniture, books, plate, pictures etc.

In spite of all these advantages, gas made from whale oil was soon bested by that made from coal, and in any case there is considerable evidence that the real reason for the decline in the Greenland fishery was that it was becoming increasingly difficult to make a go of whaling there.

The whales were becoming scarce, and those that remained were too hard to catch. Indeed, Manby sailed on Captain Scoresby's famous ship, *Baffin*, with the express purpose of testing a newly invented harpoon gun. Whalers at the time scorned firepower, preferring the hand-held harpoon. Early harpoon guns were not all that effective, and Scoresby did not use them much, but he did express a guarded approval of the new weapon. He correctly foresaw that when it came to

70

"attacking wicked fish, fish at the edge of packs, finners, razorbacks, etc., these destructive implements might be of uncommon service". Indeed they were, but I am getting ahead of myself.

All the whaling of the Western world had its origins with the Basque whalers; they went up to the north and across to the New World, and their expertise enabled other nations to take up whaling. But at the same time that the Europeans were beginning their struggle against the Greenland Right whale the Japanese, on the other side of the world, had a well-organised coastal fishery similar in many respects to that of the Basques.

At the start of the seventeenth century the Wada family controlled whaling from their base at Daichi on Kyushu, the most southerly of the four main islands of Japan. The method was essentially that of coastal whalers everywhere, although with a slightly more definite demarcation of roles than in the Bay of Biscay. Some boats specialised in harpooning, others in towing the dead whales back to shore for processing. By the end of the century, however, a very odd new method had been perfected—whaling by net.

Whales were caught on their migrations. Those going north to feed in spring were *agari kujira*, while during the autumn migration south they were *sagari kujira*. During the season, lookouts were posted on vantage points along the coast. These *yamani*—the word means "vigil on the hill"—ate a special diet said to improve their vision and powers of concentration. When a hill watcher spotted a blow he let off a series of signals, flags and rockets, to indicate the species, its position, and its route. Once the signals had been received the fleet put to sea.

It was a large collection of vessels, each with a specific task. Twenty catcher boats, each 42½ feet long and 7¼ feet across, carried fifteen crew: a headsman, thirteen oarsmen, and a boy. Eight oars, sometimes assisted by a small sail, powered each catcher towards the whale. Coming along more slowly behind the catchers were the net-carrying boats. These too had eight oars, but only ten crewmen, which with their burden of nets made these boats slower. The catchers made their way to the whale as quickly as possible, and then surrounded it. Like a flock of sheepdogs harrying a single monstrous sheep they steered the whale back towards the oncoming net boats.

The net boats took over, surrounding the whale and dropping their nets over it. This entangled the whale, making it harder for it to swim and dive, and easier for the catcher boats to move in. They clustered round, the men attempting to wound the whale with their spears and lances. The idea was not to kill the whale, but simply to weaken it as quickly as possible. Then came a moment of high drama. One of the whalemen, a specialist, jumped in the water and clambered onto the whale's vast head.

His task was to cut a hole through the tough skin between the two nostrils of the whale's blowhole. The hole was to take a rope, but the entire enterprise sounds like sheer folly. The whale might well be weakened from its many stab wounds, and encumbered by the net, but it was still capable of suddenly sounding. If this should happen, the experts advised, the hole-cutter's safest move would be to take a deep breath and hold on. At least that way he would be out of the way of the whale's flailing flukes and fins.

Somehow the hole was cut, and a rope passed through it. Now the whalers lashed their quarry to two boats, one on either side. The crew climbed into other boats, and with the whale tied fast to the boats the seamen tried to get at its heart with long swords. Violent flurries often ensued, but the whalers simply jumped overboard, clambering back when the whale had quietened down a bit. When, finally, the monster was dead the third and final element of the fleet performed its allotted task. Four towing boats set to and hauled the whale, still lashed to the two floats, back to a factory on land, where the animal was butchered.

The catch was not huge, an average of thirteen a year at one station and twenty at another, and the Japanese were much more interested in whale meat than whale oil. Cookery books of the period give many recipes for whale meat, and the eyes and mammary glands were apparently great delicacies. The average Japanese, like the average European, thought of whales as fish, an "error" that suited the Buddhists in particular. They considered meat abhorrent, but whales were fish, not flesh, and so good Buddhists ate whale meat freely. Japanese naturalists knew the score: a classification published in 1758 correctly placed whales among the mammals. The public, however, did not want to know. Food in Japan was scarce enough, and whales were too good a source to ignore.[8]

Hard though it may be to believe, the Japanese retiarii took not only Right whales with this method, but also succeeded in subduing rorquals, something the European whalers had yet to achieve. The only whale they did not really go for was the Sperm whale, because its meat was not so tasty. When, however, the Norwegians eventually invented the tools that allowed them to take Fins and Blues, the Japanese were quick to abandon their old ways in favour of the more modern style.

While the Japanese continued with their traditional hunt, more akin to subsistence than commercial whaling, the European whalers were being eclipsed by the Americans. Indeed, at one time the American whale fishery was easily the largest in the world.

It began, like the Basque fishery, as a coastal and inshore activity, more or less by accident. The whales were certainly plentiful. In 1614 Captain John Smith found so many whales along the coast of New England that he abandoned his voyage of discovery to hunt them. And the Reverend Richard Mather, who travelled to the colony in Massachusetts Bay in 1635, saw "mighty whales spewing up water in the air like the smoke of a chimney, and making the sea about them white and hoary, as is said in Job, of such incredible bigness that I will never wonder that the body of Jonas could be in the belly of a whale".[9]

At the outset the colonists took only drift whales, dead animals that had been washed ashore, and although we do not know when they first set out in active pursuit the hunt was certainly well organised by the end of the seventeenth century. In 1688 the British Agent in the colony wrote home to the Lords of Trade in England that "New Plimouth Colony have great profit by whale killing. I believe it will be one of our best returns, now beaver and peltry fayle us."[10] The earliest whalers

---

[8] It is easy to feel smug about the pragmatic Japanese attitude to the taxonomic status of whales; in Europe of the fourteenth century, and beyond, good Christians regularly ate barnacle geese on Fridays and during Lent, conveniently believing them to be derived from fish, not fowl.
[9] Dow (1985) p 6.
[10] Dow (1985) p 10.

set out from the eastern end of Long Island in boats fitted out for voyages of a couple of weeks. They did not venture far from land, generally returning to the coast each night to camp on shore.

The other main whaling post before 1700 was Nantucket. The name is now so firmly embedded in whaling's history that it is almost synonymous with the hunting of the whale. Nantucket is an island off the tip of Massachusetts, where whales were so plentiful and fearless that in the earliest days of the industry there they often came right into the harbour. They were so numerous around Nantucket that all the oil needed could be had without the ships ever venturing out of sight of land, and in the early years of the eighteenth century Nantucket quickly became the biggest whaling operation.

Lookouts on land kept watch for whales, and the boats put out when one was spotted. The carcass was towed back to shore, where the blubber was processed in the trying-out works. Right whales, however, quickly became scarce, and the industry almost died soon after its birth. After 1712, however, when the first Sperm whale was taken, business boomed.

It was all a bit of an accident. Christopher Hussey, a Nantucket whaler, was cruising offshore for Right whales when a strong northerly blew up. It carried him some way from the island, where he fell in with a school of Sperm whales. Hussey naturally killed one, and brought its blubber home, where the people of Nantucket quickly discovered that Sperm oil was better than any other. They immediately started to build bigger, sturdier ships specifically for the capture of Sperm whales. Cruises now lasted some six weeks, and the blubber was stripped from the whale at sea, being stored in casks for further processing back in port. By 1730 the shore fishery was beginning to fail, as whales near the coast became scarcer and scarcer, thanks to overfishing. Sperm oil, however, was so superior to all other oils that despite the loss of inshore whales the industry continued to boom. Larger vessels made longer voyages to more distant whaling grounds.

They generally went south, into the Atlantic, chasing after Sperm whales, and gradually ventured further afield. The coast of the Carolinas, the Bahamas, and the West Indies and Gulf of Mexico were all exploited in turn. Then across to the Azores, where Portuguese harpooners got involved, and on to the Cape

Verde islands and the coast of Africa. The spread was slow and gradual, but by 1763 American whalers were off the coast of Guinea, while a decade later, in 1774, they were taking Sperm whales in the waters off Brazil.

Whaling was so successful that it more than met the domestic requirements of the colonies, and soon an export trade was thriving. By 1730 there was regular trade in whale oil and whalebone to England and to British ports in the West Indies, and whaling was the main industry of New England. One chronicler of Provincetown complained in 1737 that "so many men are going on these voyages that not more than twelve or fourteen men will be left at home".[11]

All this was fine for the colonists, but the British government was caught with one foot in the boat and the other on shore. At home the whalers were having enormous problems in competing with the Dutch in the Greenland fishery, while in the colonies the whalers were doing extremely well and exporting their produce to England, where it added to the troubles caused by the Dutch. Parliament did not offer the bounty to whalers in the colonies, and further imposed a duty on all whale products exported to England from the colonies. Then the colonists were forbidden to send their produce elsewhere, which practically forced them to pay the duty. American and London merchants protested, but without real effect.

This was the position just prior to the American Revolution, and at that time the whale fishery was very prosperous. Between 1771 and 1775 the annual production was estimated at 45,000 barrels of Sperm oil, 8,500 barrels of oil from other whales, chiefly Right whales, and 75,000 pounds of whalebone. Sperm oil fetched £40 a ton, spermaceti £50 a ton, ordinary oil £25 a ton and whalebone about £330 a tonne. Generally speaking, one ton of whale oil fills six barrels, so at a conservative estimate the enterprise was worth a little under £500,000 a year, and a pound then was worth about £35 today.

The revolution put a stop to all that. The British government attempted to "starve New England" with embargoes on fishing and a ban on all trade in whale products. Only the people of Nantucket carried on whaling, for it was the only job they could do. In the early days of the War of Independence the British

[11] Jenkins (1921) p 226.

made a point of capturing and burning whale ships, and destroyed property all along the New England coast. So severe were the effects at Nantucket that in 1781 the British Admiral granted the Nantucketers special permission to use twenty-four vessels without fear of attacks by the British cruisers.

The end of the war found the whaling industry dead, even in Nantucket. At the outbreak of war there had been 150 ships whaling out of Nantucket. In 1784 only two or three remained. The English had captured or destroyed 134, and the rest had been lost by shipwreck. Recovery came slowly. Thanks to the enforced halt, the whales had become less shy and now were more easily killed. Whale products fetched good prices for a few years after the war, but the British market was effectively closed by the import duty, and prices soon tumbled. Most of the towns that had been involved in whaling before the war pulled out, and only Nantucket was left. The State of Massachusetts, like the British government, offered a bounty but instead of being based on tonnage it was based on the amount of oil brought back into the state.

Between 1776 and 1815—the War of Independence and the War of 1812—whaling's fortunes fluctuated. Trade with France increased briefly but was then curtailed by the revolution there. At home the general increase in prosperity brought about an increase in the demand for whale oil, and Sperm candles in preference to those made of tallow. And the export trade, for example to the thriving new colonies in the West Indies, also increased.

By 1820 the whalers were regularly venturing into the Pacific, and between then and 1835 the industry grew steadily. The twenty years after 1835 represent the zenith of the Yankee whaling industry. The fleets roamed the oceans of the world in search of Sperm whales, and discovered virgin hunting grounds for Right whales in the Arctic and Antarctic. New uses had been found for whale products. The bone was finding its way into all manner of domestic goods where springiness was needed, and the oil enjoyed steady and increasing demand for lighting, both as Sperm candles and in whale-oil lamps.

It was the opening of the first petroleum well in Pennsylvania in 1859 that closed the American Sperm oil industry. The struggle was short and sharp. Kerosene rapidly found favour as an illuminant, mostly on grounds of economy. After refining the kerosene out of raw petroleum the residual oils made excellent lubricants. And the wax, or paraffin, went into candles. Thus the three primary uses of whale oil were fulfilled, more cheaply and less dangerously, by petroleum.

Before this, the industry had been huge. Some 70,000 people were employed throughout the business of whaling, which was estimated to be worth about $70 million (£14 million). The fleet averaged 600 vessels, worth about $21 million (£4.3 million). It brought in whale products worth about $8 million a year (£1.64 million), and a pound then was worth 26 of today's.

The profits to be had were quite enormous. As an example, consider the *Lagoda*, a whaler from New Bedford, commissioned at the height of the boom and sold only when the slump had unmistakably set in. The *Lagoda* made twelve voyages between October 1841 and July 1886. Ten of these resulted in a profit and only two, the tenth and the twelfth, in a loss. The net gain to the owners during that time was some $652,000, but a better grasp of the profitability of the enterprise can be gained from looking at the dividends. Expressed in percentages, the dividends for the twelve voyages were: 29.6, 120.5, 66.9, 177.2, 100, 96.9, 363.5, 219.0, 115.2, loss, about 10, loss. This is an average, for the profitable voyages, of almost 130 per cent. Of course some of the voyages were long ones, and the owners twice sustained a loss, but even so the exceptional seventh voyage, which lasted forty-four months, provided an average monthly return of 8.25 per cent. It was generally reckoned that a single good-sized whale would finance the entire whaling voyage.

It wasn't all profit, however. Luck played an enormous part in the business. There were huge gains: in 1866, for example, two whalers from New Bedford each made a profit of $125,000 (500 per cent) on capital employed of $25,000. There were also huge losses. In 1858, sixty-eight vessels returned to New Bedford and Fairhaven, but forty-four of them had made a loss and the total loss was about one million dollars. In 1871 the entire Arctic fleet was destroyed by pack ice; thirty-four vessels were written off, and the total loss was more than $2 million.

Petroleum conquered whale oil slowly, but the Civil War had a more immediate effect. Confederate privateers played havoc with the Yankee whalers. The *Shenandoah* went up into the

Bering Sea, where she captured twenty-nine whalers, burning twenty-five and appropriating the other four as transports. The Yankees lost fifty whalers in the war, and the government purchased another forty to create the Charleston stone fleet, which was sunk in an attempt to blockade Charleston harbour.

After the Civil War there was a partial revival of whaling, mostly based in San Francisco. Oil was no longer so valuable, but the price of whalebone kept rising. Gradually the whaling owners shifted their operations from the New England ports of New Bedford, Sag Harbor and Nantucket to the west coast. Steam helped enormously. The English had first used steam in whalers in 1857, but the Americans did not adopt it until 1880. When they did, it revolutionised Arctic whaling. Before steam, the fleet had spent the winter in San Francisco or some other Pacific port, refitting and perhaps whaling in the calving lagoons of Baja California. They then sailed north in spring, waiting for the ice to break up so that they could get through the Bering Strait and into the Beaufort Sea. With steam power, it became customary for the whalers to spend the winter in the Arctic, so as to be first in the field when the ice broke up and the whales returned.

There were some curious influences on the development of whaling in America. One was the discovery of gold in California in 1849. During the gold rush whole crews would desert when the ship came in to one of the Pacific ports to refit. In fact, signing on for a whaler became recognised as one of the cheapest ways for young men to get from the east coast to the goldfields, and the capitalists of whaling lost large amounts when their ships were laid up for want of a crew. Another was the manufacture of cotton goods, which began in New Bedford in 1846. The profits were not so spectacular, but they were steadier, and there is no doubt that capital moved out of whaling and into cotton. Even so, the Yankee whalers prospered long, and they prospered well, primarily on the backs of the mighty Sperm whales.

If there is one name associated with the Yankee Sperm whalers, it is Herman Melville's. *Moby-Dick*, subtitled simply *The Whale*, is a superb novel. It is also an extremely accurate account of the business of whaling. I can do no better than to quote Melville's own words: "It was my turn to stand at the foremast-head," says the man who asks us to call him Ishmael:

and with my shoulders leaning against the slackened royal shrouds, to and fro I idly swayed in what seemed an enchanted air. No resolution could withstand it; in that dreamy mood losing all consciousness, at last my soul went out of my body; though my body still continued to sway as a pendulum will, long after the power which first moved it is withdrawn.

Ere forgetfulness altogether came over me, I had noticed that the seamen at the main and mizen mast-heads were already drowsy. So that at last all three of us lifelessly swung from the spars, and for every swing that we made there was a nod from the slumbering helmsman. The waves, too, nodded their indolent crests; and across the wide trance of the sea, east nodded to west, and the sun over all.

Suddenly bubbles seemed bursting beneath my closed eyes; like vices my hands grasped the shrouds; some invisible, gracious agency preserved me; with a shock I came back to life. And lo! close under our lee, not forty fathoms off, a gigantic Sperm Whale lay rolling in the water like the capsized hull of a frigate, his broad, glossy back, of an Ethiopian hue, glistening in the sun's rays like a mirror. But lazily undulating in the trough of the sea, and ever and anon tranquilly spouting his vapory jet, the whale looked like a portly burgher smoking his pipe of a warm afternoon. But that pipe, poor whale, was thy last. As if struck by some enchanter's wand, the sleepy ship and every sleeper in it all at once started in wakefulness; and more than a score of voices from all parts of the vessel, simultaneously with the three notes from aloft, shouted forth the accustomed cry, as the great fish slowly and regularly spouted the sparkling brine into the air.

"Clear away the boats! Luff!" cried Ahab. And obeying his own order, he dashed the helm down before the helmsman could handle the spokes.

The sudden exclamations of the crew must have alarmed the whale; and ere the boats were down, majestically turning, he swam away to the leeward, but with such a steady tranquility, and making so few ripples as he swam, that thinking after all he might not as yet be alarmed, Ahab gave orders that not an oar should be used, and no man must speak but in whispers. So seated like Ontario Indians on the gunwales of the boats, we swiftly but silently paddled along; the calm not admitting of the noiseless sails being set. Presently, as we thus glided in chase, the monster perpendicularly lifted his tail forty feet into the air, and then sank out of sight like a tower swallowed up.

"There go flukes!" was the cry, an announcement immediately followed by Stubb's producing his match and igniting his pipe, for now a respite was granted. And after the full interval of his sounding had elapsed, the whale rose again, and being now in advance of the smoker's boat, and much nearer to it than to any of the others, Stubb counted upon the honor of the capture. It was obvious, now, that the whale had at length become aware of his pursuers. All silence or cautiousness was therefore no longer of use. Paddles were dropped, and oars came loudly into play. And still puffing at his pipe, Stubb cheered on his crew to the assault.

Yes, a mighty change had come over the fish. All alive to his jeopardy, he was going "head out"; that part obliquely projecting from the mad yeast which he brewed.

"Start her, start her, my men! Don't hurry yourselves; take plenty of time—but start her; start her like thunder-claps, that's all," cried Stubb, spluttering out the smoke as he spoke. "Start her, now; give 'em the long and the strong stroke, Tashtego. Start her, Tash, my boy—start her, all; but keep cool, keep cool—cucumbers is the word—easy, easy—only start her like grim death and grinning devils, and raise the buried dead perpendicular out of their graves, boys—that's all. Start her!"

"Woo-hoo! Wa-hee!" screamed the Gay-Header in reply, raising some old war-whoop to the skies; as every oarsman in the strained boat involuntarily bounced forward with the one tremendous leading stroke which the eager Indian gave.

But his wild screams were answered by others quite as wild. "Kee-hee! Kee-hee!" yelled Daggoo, straining forwards and backwards on his seat, like a pacing tiger in his cage.

"Ka-la! Koo-loo!" howled Queequeg, as if smacking his lips over a mouthful of Grenadier's steak. And thus with oars and yells the keels cut the sea. Meanwhile, Stubb, retaining his place in the van, still encouraged his men to the onset, all the while puffing the smoke from his mouth. Like desperadoes they tugged and they strained, till the welcome cry was heard—"Stand up, Tashtego!—give it to him!" The harpoon was hurled. "Stern all!" The oarsmen backed water; the same moment something went hot and hissing along every one of their wrists. It was the magical line. An instant before, Stubb had swiftly caught two additional turns with it around the loggerhead, whence, by reason of its increased rapid circlings, a hempen blue smoke now jetted up and mingled with the steady fumes from his pipe. As the line passed round and round the loggerhead; so also, just before reaching that point, it blisteringly passed through both of Stubb's hands, from which the hand-cloths, or squares of quilted canvas sometimes worn at these times, had accidentally dropped. It was like holding an enemy's sharp two-edged sword by the blade, and that enemy all the time striving to wrest it out of your clutch.

"Wet the line! wet the line!" cried Stubb to the tub oarsman (him seated by the tub) who, snatching off his hat, dashed the sea-water into it. More turns were taken, so that the line began holding its place. The boat now flew through the boiling water like a shark all fins. Stubb and Tashtego here changed places—stem for stern—a staggering business truly in that rocking commotion.

From the vibrating line extending the entire length of the upper part of the boat, and from its now being more tight than a harpstring, you would have thought the craft had two keels—one cleaving the water, the other the air—as the boat churned on through both opposing elements at once. A continual cascade played at the bows; a ceaseless whirling eddy in her wake; and, at the slightest motion from within, even but of a little finger, the vibrating, cracking craft canted over her spasmodic gunwale into the sea. Thus they rushed; each man with might and main clinging to his seat, to prevent being tossed to the foam; and the tall form of Tashtego at the steering oar crouching almost double, in order to bring down his centre of gravity. Whole Atlantics and Pacifics seemed passed as they shot on their way, till at last the whale somewhat slackened his flight.

"Haul in—haul in!" cried Stubb to the bowsman; and

facing round towards the whale, all hands began pulling the boat up to him, while yet the boat was being towed on. Soon ranging up by his flank, Stubb, firmly planting his knee in the clumsy cleat, darted dart after dart into the flying fish; at the word of command, the boat alternately sterning out of the way of the whale's horrible wallow, and then ranging up for another fling.

The red tide now poured from all sides of the monster like brooks down a hill. His tormented body rolled not in brine but in blood, which bubbled and seethed for furlongs behind in their wake. The slanting sun playing upon this crimson pond in the sea, sent back its reflection into every face, so that they all glowed to each other like red men. And all the while, jet after jet of white smoke was agonizingly shot from the spiracle of the whale, and vehement puff after puff from the mouth of the excited headsman; as at every dart, hauling in upon his crooked lance (by the line attached to it), Stubb straightened it again and again, by a few rapid blows against the gunwale, then again and again sent it into the whale.

"Pull up—pull up!" he now cried to the bowsman, as the waning whale relaxed in his wrath. "Pull up—close to!" and the boat ranged along the fish's flank. When reaching far over the bow, Stubb slowly churned his long sharp lance into the fish, as if cautiously seeking to feel after some gold watch that the whale might have swallowed, and which he was fearful of breaking ere he could hook it out. But that gold watch he sought was the innermost life of the fish. And now it is struck; for, starting from his trance into the unspeakable thing called his "flurry", the monster horribly wallowed in his blood, overwrapped himself in impenetrable, mad, boiling spray, so that the imperilled craft, instantly dropping astern, had much ado blindly to struggle out from the phrensied twilight into the clear air of the day.

And now, abating in his flurry, the whale once more rolled out into view; surging from side to side; spasmodically dilating and contracting his spout-hole, with sharp, cracking, agonized respirations. At last, gush after gush of clotted red gore, as if it had been the purple lees of red wine, shot into the frighted air; and falling back again, ran dripping down his motionless flanks into the sea. His heart had burst!

"He's dead, Mr Stubb," said Tashtego.

"Yes; both pipes smoked out!" and withdrawing his own from his mouth, Stubb scattered the dead ashes over the water; and, for a moment, stood thoughtfully eyeing the vast corpse he had made."[12]

There is, of course, much more to whaling than the capture, and Melville deals faithfully with all of it: the butchery, severing the whale's great head from its body; the cutting in, removing the thick blanket of blubber; trying out, boiling the oil from the blubber; and all the other thousands of tasks that made up an average whaling voyage. The cruise of the *Pequod* was, I admit, no ordinary whaling voyage. Most whaling captains did not pursue a single beast so single-mindedly. But *Moby-Dick*, quite apart from being an exceptionally good tale, is also an accurate portrait of Yankee whaling at its best.

Not surprisingly, for Melville had shipped aboard a whaler, the *Acushnet*, an experience that provided him with many of the stories which later made him such a name. But the story of the begetting of *Moby-Dick* is a strange tale itself. Six months out of Fairhaven, Massachusetts, the *Acushnet* was in the South Pacific when she came alongside another whaler, the *Lima*, from Nantucket. As was customary, the two ships came together for what the whalers called a gam; the men of the *Acushnet* swept over to the *Lima* to swap stories and plugs of tobacco, while the officers gathered on the *Acushnet*. The two vessels were close to the middle of the most open stretch of the Pacific. There were few whales around, and plenty of time for conversation and telling tales.

Aboard the *Lima* was a young foremast hand named William Chase, and he certainly had a tale to tell. He was the son of Owen Chase, who had been the first mate on a Nantucket whaler called the *Essex*. On 20 November, 1820, the *Essex* was rolling along the equator, near to where the *Acushnet* and the *Lima* now lay together. The *Essex* gave chase to a large school of Sperm whales. Down went the three whaleboats, with Chase in command of one of them. Halfway to the school the whales reacted, diving for the bottom. The ocean surface fell quiet, and Chase and his men rested on their oars. Chase moved forward, handing the steering oar to one of the crew and taking up the big harpoon. Unlike so many other harpooners, Chase had no

[12] Melville (1972) pp 388–93.

patience with the precarious scramble when harpooner and headsman changed places. He did both jobs.[13]

Still the water remained smooth, as if no whale had ever been there. Then, with almost no warning, a medium-sized whale breached the surface just a few yards from the boat. Chase hurled the harpoon with all his power. He heard the satisfying thud as the iron head of the harpoon pierced the whale's skin, and watched as the smoking line was pulled out of the boat. And then, just as the crew had shipped oars and braced themselves for the tumultuous sleighride, came a smash. The whale's flukes struck the side of the boat. They crashed through the sides like a fish through wet newsprint, and as the whale began to run water poured into the boat.

Chase hardly hesitated. He grabbed the axe that every boat carried for occasions such as this, and cleaved the line. The whale swam off, the harpoon still in its back, while the crew stuffed their clothes into the gaping hole to staunch the flow and rowed as hard as they could for the *Essex*. Chase took command, and headed in the direction of the other whaleboats. As they sailed Chase supervised the patching of his whaleboat, thinking he could lower away and be of more help to the others. He looked up, and saw a whale break the surface just off the bows. It was eighty-five feet long, and spouted quietly a couple of times. Chase ignored it. He had more important work to do than gaze at whales.

When he glanced up again he was stunned by what he saw. The whale turned, aimed deliberately at the *Essex*, and began to come straight for the whaler with astonishing speed and power. There was a thunderous crack as whale hit whaler. The men were thrown to the decks, and moments later they could plainly hear water streaming through the hole into the hold. The whale made off, apparently more maddened than hurt by the blow, and swung round for a second attack. Again the sickening impact, and the flow of the sea into the *Essex* redoubled.

Chase ordered a whaleboat prepared and his crew scurried about loading vital food and equipment into her. By this time the sea lapped at the knees of the men on deck. They jumped into the whaleboat and pushed off, just as the *Essex* rolled beneath them, kept afloat on her side by a little air trapped in

---

[13] Melville felt the same way; chapter 62 of Moby-Dick is devoted to just this point.

the holds. By now the two other whaleboats had turned back to the *Essex*, and when they reached the wreck the men broke through the planking to rescue what supplies they could. Then the three boats made fast to the hulk and spent the night.

Next day the captain called a conference. The islands of the Marquesas lay 1,400 miles southwest of them, but the officers decided instead to head south and east, hoping to find some unknown Pacific island or else make landfall on the coast of South America, 3,000 miles away. The decision to adopt the course they did proved ironic in the extreme.

The islands of the Marquesas were inhabited, and the natives were generally believed to be cannibals. That is why the captain of the *Essex* chose to head southeast. Cannibalism did play a part in the wreck of the *Essex*, but not the part everyone imagined it might.

After more than eight weeks at sea Chase emerged from a sudden storm to find his boat separated from the other two. The five men somehow carried on, some rations left, which Chase kept in a locked chest that he guarded at night with a pistol. One of the men died, and his fellows slipped him into the sea. By 27 January, the sixty-ninth day of the ordeal, the men were so weak that nobody could raise the tattered makeshift sail, or even hold the boat on course. But it was not until day eighty-one that the first man went mad.

Isaac Cole sat up and demanded a cup of water and a napkin. He kept this up for an hour or more, as his companions watched in silence. Each presumably wondered whether he might not be next. Cole died that afternoon. There were three days' biscuits left. Within one or two days the next seaman would probably follow Cole. But there was a way to keep alive, perhaps for a couple of weeks. Nobody said anything until next morning, when Chase himself suggested that they eat Cole, and perhaps save themselves.

"We set to work," Chase recounted, "as fast as we were able to . . . We separated his limbs from his body and cut all the flesh from the bones, after which we opened the body, took out the heart, then closed it again—sewed it up as decently as we could and committed it to the sea."[14]

Cannibalism, far from ending the men's lives, saved them.

[14] Whipple (1973) p 36.

On the ninety-first day, just nine days after the three remaining sailors had eaten their companion, they were rescued by the brig *Indian* from London. In their 27-foot whaleboat Chase and his two remaining fellows had drifted 4,500 miles.

One of the other boats was never seen again but the third, incredibly, was also saved. The captain's boat, like the mate's, turned to cannibalism, but with an even crueller twist. Two men died in quick succession, and after some soul-searching were eaten. They restored Captain Pollard and the three other men to sufficient health that there was little prospect of another meal of human flesh, at least for some while. They drew lots. The captain's nephew Owen Coffin, cabin boy on the *Essex*, chose to die and one Charles Ramsdell to kill him.

Ramsdell begged young Coffin to trade places, but Coffin insisted that it was his right to die for the remaining three. The cabin boy's body kept them alive for a further ten days and then the third man died. Pollard and Ramsdell shared his flesh, and were again starving when, on the ninety-sixth day of their ordeal, they too were picked up and rescued.

Twenty-eight men put to sea after the wreck of the *Essex*. Eight of them survived. Owen Chase, when he had recovered somewhat, wrote down his account of the ordeal. William Chase had a copy of his father Owen's *Narrative of the Most Extraordinary and Distressing Ship Wreck of the Whale-Ship Essex of Nantucket; Which Was Attacked and Finally Destroyed by a Large Spermaceti-Whale in the Pacific Ocean* in his sea-chest on the *Lima*. William Chase gave it to Melville, who read it and returned it the next morning. Owen Chase's story became *Moby-Dick*, and years after this chance meeting, long after he had written his novel, Melville recalled it clearly. "The reading of this wondrous story upon the landless sea, & close to the very latitude of the shipwreck had a surprising effect on me."[15]

Melville's meeting with young Chase, and his reading of the wreck of the *Essex*, sowed the seed of *Moby-Dick*, but the story of this greatest whaling story does not end there. Nine months after the meeting between Melville's *Acushnet* and William Chase's *Lima* the *Acushnet* made another chance encounter. This was the *Charles Carroll*, another Nantucket whaler. Again, the two ships came together for a gam, and Melville watched as the

[15] Melville (1972) p 18.

captain of the *Charles Carroll* came aboard to visit the officers of the *Acushnet*. Whispers coursed between the sailors. It was Owen Chase, first mate of the *Essex*. Years later, Melville wrote about that second fateful meeting with Chase:

> [S]o it came to pass that I saw him. He was a large, powerful well-made man; rather tall; to all appearances something past forty-five or so; with a handsome face for a Yankee, & expressive of great uprightness & calm unostentatious courage. His whole appearance impressed me pleasurably. He was the most prepossessing-looking whale-hunter I think I ever saw.
>
> Being a mere foremast hand I had no opportunity of conversing with Owen (tho' he was on board our ship for two hours at a time) nor have I seen him since.[16]

In fact, Melville had not seen Chase at all, although it was an understandable error. Chase *had* been master of the *Charles Carroll* on its previous voyage, but in the spring of 1842 he was at home on Nantucket, involved in the almost unheard of activity of suing his wife for divorce on the grounds of adultery. The tall prepossessing-looking whale-hunter in command of the *Charles Carroll* at the time was someone else entirely, a man called Thomas Andrews. Of course, it does not matter whether it was Chase or Andrews that Melville saw. "What mattered," as A. B. C. Whipple has written, "was that Melville thought this was Chase, and what the incident proved was that the story of the *Essex* had by now got firm hold of the young man's imagination".[17]

There was yet another source that fuelled the growth of *Moby-Dick*; the real rogue whale. There were many tales of whales that turned on their tormentors, and not a few of a white whale. Some said that this beast was pure white, white as snow, the strange, vivid eerie whiteness that Melville regarded as so symbolic and so shocking. Others said he was off-white—creamy—and yet others that he was grey with a white patch on his looming, bulbous head. They called him Mocha Dick, because he had first been spotted near the island of Mocha off the Chile coast. Quite apart from his colour, what made Mocha

[16] Melville (1972) p 17.
[17] Whipple (1973) p 45.

Dick out of the ordinary was his ferocity; other Sperm whales were content to swim along, at least until harpooned. Then of course they might unleash their power on the puny whaleboats. Mocha Dick, however, turned the tables. He apparently cruised the Pacific, hunting down whaleboats and getting his retaliation in first.

The legend of Mocha Dick began quietly enough, a single incident in which a mad whale attacked a whaleboat. The whale that stoved in the *Essex* was not, according to Chase, white. But more than a decade after the *Essex* went down similar stories began to crop up everywhere. What had been a rarity became almost commonplace, and Mocha Dick became the main topic of conversation whenever whalemen gathered to chat. His tally—of boats smashed like kindling and men carried off—grew monthly. Perhaps all the incidents blamed on Mocha Dick did happen. Perhaps there were even more, which went unremarked. But if there was a whale called Mocha Dick he could not have done all the dark deeds ascribed to him. Not even a monstrous white whale could cover the distances Mocha Dick was required to by the logbook tales. Nevertheless, Mocha Dick became *the* rogue whale, responsible for scores of infamous misdeeds.

In June 1840 the English whaler *Desmond* was cruising 215 miles off the coast of Valparaiso. A lone Sperm whale, bigger than anyone on board had ever seen, breached two miles off. The boats were lowered and gave chase, but before they got near he turned and came for them. He was more grey than black, with a pure white scar across his head. The boats tried to flee, but the whale caught the leading one, struck it head on, and sent the crew flying. Then he crunched the splintered boat. The second boat picked up the men from the first, when the great whale burst from the water below it. That boat too was smashed. Mocha Dick apparently surveyed the wreckage before he swam off, and when the *Desmond* picked up her men two were missing.

A month later, 500 miles away, the Russian barque *Serepta* finished off a lone whale. As the boats were towing it back to the ship Mocha Dick appeared. He munched one boat to matchwood and then made for the second. The mate cunningly hid his boat behind the dead carcass, temporarily foiling the rogue whale. The men rowed to the ship for their lives, and made it, but could not retrieve their prize. Mocha Dick guarded the dead whale until the Russians gave up and the *Serepta* sailed away.

The skipper of the Bristol whaler *John Day* vowed that if he ever came across the monster he would kill him, or lose every man and boat in the attempt. In May 1841 he was in the South Atlantic when Mocha Dick suddenly breached, less than 300 yards away. The captain lowered three boats and went after the rogue. A mate more skilful than the others moved the boat aside just as the monster came for them, and managed to plunge a harpoon into his side. The crazed whale towed the boat at breakneck speed for about three miles, then suddenly reversed and came straight for them. Mocha Dick hit broadside, powered on over the boat, and beat it with his enormous tail flukes. Two of the crew had vanished. Mocha Dick hove to and waited.

The two remaining boats took up the chase. One got hold of the first's line, which was still embedded in the white whale's side. The monster sounded again, coming up with a whoosh directly below the third whaleboat. The 27-foot craft flipped end over end through the air, but miraculously nobody was killed. The captain of the *John Day*, however, had had enough. He ordered his crew back and they sailed away, leaving Mocha Dick snorting and thrashing his tail.

There were many other such incidents, from the South Atlantic, across the Pacific and right up to the coast of Japan. In all likelihood they were not all the handiwork of a single whale, though whalers maintained that they were. After two decades of reports about a rogue white Sperm whale, the stories suddenly dry up, which suggests that perhaps there really was only one whale. In any case, it hardly matters, for Mocha Dick was a major inspiration for *Moby-Dick*.

The book had just been published, in November 1851, when news reached Melville of a catastrophe that had befallen a whaler called the *Ann Alexander*. A maddened whale had smashed the boat and, like the men of the *Essex* thirty years before, the crew had taken to their boats. The whaling grounds, however, were now crowded with adventuring vessels, and the survivors were picked up after just one day adrift, compared to Chase's ninety-one. When Melville heard about the *Ann Alexander* he replied to the friend who had brought the episode to his attention: "I make no doubt it *is* Moby Dick himself . . . Ye Gods! what a Commentator is this *Ann Alexander* whale. I wonder," he added, "if my evil art has raised the monster?"[18]

Melville's whalers were men of sail, but steam was soon to replace them. The American whalers used steam to great effect in the Arctic, and in England preparations were made to adapt the American experience, creating ships that would go after whales and seals in the Arctic ice. Jenkins describes one such vessel:

> One of the Peterhead whalers attracted much attention. The *Empress of India*, built of iron, was specially fitted out for the trade. She was strongly fortified, being twelve feet thick forward, and carried eleven boats. The bottom of the captain's gig was bronze. No expense was spared in her outfit, her crew consisting of one hundred and ten men. All the crew expected to make a small fortune, and looked on the old sailers with contempt. Some of the officers were so sure of getting full of seals that they made all their plans for the future; they were going to fall in with the north end of the main body of seals and sweep through the centre, leaving the rest for those who were fortunate enough to be in their company. However, the first piece of heavy ice penetrated their port bow, and they foundered in four hours, all hands being saved by the despised sailers.[19]

At first steam was not successful. Many steamships came back damaged, and Barron, master of the *Truelove*, proclaimed that no matter how strongly built, iron vessels could not cope with Greenland pack ice. A few years later, he had changed his mind. In 1861, Barron wrote that "this year would prove the death-blow to sailing vessels. Men having experienced the great difference between steam and sail, few will go hereafter in a sailing ship if they can get into a steamer."[20] The problem, for ships under sail, was the drift ice. Often this was only a temporary barrier, with clear water beyond, but while it took sailing ships enormous efforts to cross the ice, if indeed they

---

[18] Whipple (1973) p 72.
[19] Jenkins (1921) p 257.
[20] Jenkins (1921) pp 257–8.

ever made it, steamships did so with certainty in a matter of days. The old whalers, under sail, considered themselves lucky if they got through the ice one year in three, and even then it involved between a month and sixty days of unremitting hard work. Steam whalers got through every year, often in sixty hours.

Barron changed his mind as he was trying to get the *Truelove* through the ice. "After toiling all day we only succeeded in getting a mile. The S.S. *Narwhal* came to our relief, and towed us into clear water without the least difficulty. This showed the superiority of steam over sailing vessels."[21]

It was the power of steam that resurrected the North Atlantic fishery. Whaling had been declining from about 1830, and by the 1860s there were very few whalers left. Mostly it was the Norwegians who persisted with whaling under sail, and it was the Norwegians who ushered in the final great phase of modern whaling. By 1880 the Right whale, so valuable for its whalebone, was nearly extinct, and the whalers were mostly after the Sperm whale. The rorquals—Fins, Blues and Seis—were still plentiful, but their whalebone was inferior to that of the Right whales and they were not very oily. More to the point, they were much harder to catch. They swam faster, further out at sea, and were prone to dive long and deep, all of which made capture by hand-held harpoon hazardous in the extreme. Rorquals also sank when dead, which made the few that were captured much harder to deal with. Svend Foyn and his inventions changed all that.

Svend Foyn was born at Tönsberg in Norway in 1809. He started his life as a sealer, but kept a close watch on whaling and decided that the industry might have a future, but only if there was some way of handling the great rorquals, the only whales that were still really plentiful. The Norwegians had not been one of the great whaling nations. They did take a few of the little Minke whales, doing so in an extremely curious way.

When whales had entered a fjord the people would close the mouth of the fjord with a net. They then fired a rusty and stained cross-bow bolt into the whale, preferably a bolt that had already seen much service in this particular pursuit. As if

---

[21] Jenkins (1921) p 258.

by magic, the whale weakened, and could be quite easily despatched after a few days. We now know that the whale probably succumbed to septicaemia, which is why an old and used bolt worked better. Still, if they were late coming to whaling, the Norwegians more than made up for lost time.

Foyn made money at his sealing, and the profits enabled him to carry out costly experiments on how to tackle the rorquals. In 1863 he ordered a revolutionary ship, called *Spes et Fides*. She was small, 86 tonnes, schooner rigged but with an auxiliary steam engine that gave her a speed of seven knots. Most importantly, her forecastle resembled a miniature warship, bristling with a battery of seven experimental harpoon guns.

Foyn and the *Spes et Fides* took off for the Finmark coast in the spring of 1864. They soon came upon whales, but the very first attempt was almost the last. The harpoon line coiled around Foyn's leg and hurled him overboard into icy seas. One seaman remarked, with typical Norse phlegm, "I thought the captain was drowned," but Foyn freed himself, swam back to the ship, was picked up, and resumed operations. Later that same experimental summer a harpooned whale pulled the boat for twelve hours at an estimated eighteen knots before Foyn reluctantly cut the line. 1865 was spent making modifications to the equipment, and in 1866 Foyn sailed for Iceland, where he studied the shoulder guns being used by the Americans. Another cruise the next year brought a booty of only one whale.

Foyn was a godly man, who inscribed pious biblical quotations on each page of his logbook, but he was also a wealthy one. Faith and finance kept Foyn going. "And," as Paul Budker noted, "it must be acknowledged that Heaven did favour him in several ways, particularly through one of its earthly representatives, Pastor Hans Morton Thrane Esmark."[22] Foyn was having great difficulty devising a method for reliably detonating an explosive charge inside the whale. He turned to Pastor Esmark, whose hobby was pyrotechnics, and the man of God perfected the weapon of man.

Foyn's gun harpoon consisted of three parts: the pole, attached to the line; the barbed head, attached to the pole; and the explosive charge, screwed to the head. The head carried four hinged barbs, which opened out when the harpoon penetrated

[22] Budker (1958) p 107.

the whale, making the line fast. The impact also broke a glass tube containing sulphuric acid. This was Esmark's contribution, for the acid detonated the charge, which ultimately killed the whale. Foyn's other great invention was the inflation lance, a hollow spear that the whalers used to pump air into the dead whale's body. The air kept the whale afloat, so that it could be towed back to a land station for processing.

After four years of trial and error, refinement and development, Foyn was ready for a real hunting season. The *Spes et Fides* took thirty whales in a single short season. Modern whaling had arrived. What the rorquals lacked in quality of oil was more than made up by the quantity of whales that could be taken. The same fate that had overtaken the Right whales now awaited the Seis, the Humpbacks, the Fins, and the mighty Blues.

The Norwegians quickly cleaned up the waters close to home. Their own government prohibited whaling off the Norwegian coast, the stocks having been dangerously depleted, so the Norsemen set off abroad. They moved to the Orkneys and Shetlands, off Scotland, where trouble was not long in coming. There was a clash between the whalers and the herring fishermen, who complained that the whalers were disturbing the shoals and that the processing of the whales was offensive and a threat to public health. A special committee took evidence at several places in the Shetlands and also visited the whaling stations. The committee concluded that unrestricted whaling might indeed damage the herring fishery, but that a total prohibition would have worse consequences still. It could, the committee said, lead to floating factories being set up, off the islands, and it might lead the whalers to move their base of operations to the Faroes while continuing to whale in Shetland waters. In either event, control would be even more difficult to implement. These arguments were prescient in their anticipation of later whaling methods. The committee, while recognising the dangers of a ban, did not feel that whaling ought to be completely unregulated.

> Unrestricted whaling would be an evil on other grounds than its possible danger to the herring fishery. It could not last long. The Basque and the Greenland whaling industries came to an end by the practical extermination of the species

pursued. With the means of destruction now brought to deadly perfection the same fate would overtake the Finners off our coasts in a very short time. That would be an evil in itself, and, while a few companies might go out of the business with a large profit, the local industry would be brought into being only to perish in a few years, and leave the inhabitants worse off than ever.[23]

Regulations were enacted. Whalers had to be licensed, and only British subjects would get permission. The government also set size limits, closed seasons, and the like, and these preserved the Scottish whale fishery for some time, but overfishing still depleted the stocks.

At the end of the nineteenth century whaling appeared to be dying out all around the world. In the European Arctic the hunt for the Greenland Right whale had long been abandoned. Those ships that did venture north were equipped to take seals, walruses, and any other animals that might yield fur or oil, and still voyages were not often profitable. They also went for the small white, or beluga whales, which provided not only oil but also a skin that could be made into excellent leather. Rorquals were being hunted from land stations based in Finmark, Tromsö and Iceland. The average catch was just under 2,000 animals a year. A third whale captured in European waters was the Grindhval, or Pilot whale, a very small whale hunted by islanders in the Shetlands, Orkneys, and Faroes. Trade in the Bottlenose whale began in about 1881, and grew rapidly as the excellent oil of this species found a ready market. By and large, however, whaling in northern waters was moribund.

Before the century was out the whalers had begun to cast their eyes south to the Antarctic, the last virgin field. The Tay Whale Fishing Company of Dundee sent four steamers south to the Falkland Islands in 1891, and thence to the Antarctic. They captured many seals, but no whales, and the voyage was not a financial success. Germans, based in Hamburg, sailed to Antarctica, but concerned themselves exclusively with sealing. The next attempt was Norwegian.

The *Antarctic*, a specially outfitted steamer costing £5,000, sailed south in 1893, equipped to hunt Right whales and Sperm

[23] Jenkins (1921) p 277.

whales. On the voyage south they saw many rorquals, but could not take them as their gear was not suitable. After sailing the southern oceans for two years the Norwegians concluded that there were not enough Right whales available in the Antarctic summer to make whaling there worthwhile. Half a century earlier, Sir James Clark Ross, the polar explorer, on his return to Cape Town from the Antarctic, reported seeing five to six hundred whalers, mostly American, and all were hunting Right whales in the very same waters that the Norwegians now declared unprofitable. Most of the ships that Ross saw made good profits, but such was the destruction they wrought on the stocks that just fifty years later there were not enough Right whales left to support a much more efficient industry.

The voyage of the *Antarctic* did, however, demonstrate that rorquals were present in the Southern Ocean in vast numbers. Hardly a day passed without Blues, Fins, Seis or Humpbacks being seen, and the Norwegians were quick to return to the Antarctic, this time with the correct equipment.

In 1904 the Norwegians established a company, the Sociedad Argentina de Pesca, in Buenos Aires. The capital was Argentinian, but all the personnel were Norwegian; the manager of the company was a Norwegian whaling captain, C. A. Larsen, who had sailed as a captain under the polar explorer Fridtjof Nansen. Larsen built a whaling station on the island of South Georgia, where he had a larger than average whaler, capable of towing six Blue whales. Two other vessels ferried between Buenos Aires and South Georgia, bringing provisions to the whalers and taking the oil back.

The station was extraordinarily successful, capturing a hundred rorquals in the six months fishing up to June 1905. By 1910 there were Norwegian whaling companies around the world, and they too were very successful, leading to even more expansion. In 1911, the estimate for the annual slaughter was more than 20,000 whales. By 1912 almost the entire world whaling industry was in Norwegian hands, with sixty companies at work. Life at those whaling stations is brilliantly portrayed in a memoir written by Robert Cushman Murphy.

Murphy was an American naturalist, with a degree from Brown University. In 1911, the director of the American Museum of Natural History offered him the chance to voyage to the edge of the Antarctic on a New Bedford vessel, the brig *Daisy*, which was going south to hunt for whales and seals. Murphy was reluctant, for he was engaged to be married the following June to Grace Emeline Barstow. She, however, persuaded him to accept the post, and brought their marriage forward. They were wed on 17 February, 1912, steamed down to the Lesser Antilles on 25 May, and joined the *Daisy*. On 1 July, Murphy waved goodbye to his wife of four months and the *Daisy*[24] set sail for the Southern Ocean. Murphy began a log, covering the period 1 July, 1912 to 8 May, 1913. It was subsequently edited and published as *Logbook for Grace*, and it offers a fresh, exciting glimpse of all the jewels that might attract a young naturalist's attention on a voyage to the Antarctic. Not least among these was the whaling industry of South Georgia.

Larsen created a tiny haven called Grytviken on South Georgia. Various explorers had probably visited South Georgia before, but it was Captain Cook who claimed, and named, the island for King George in January 1775. The hamlet was on the shores of the sheltered waters of Cumberland Bay, in the middle of the north coast of the island. Cumberland Bay is backed by hills that are often covered in snow. Behind the hills is a great range of mountains, perpetually white, a dramatic backdrop when the waters of the bay are smooth.

By the time Robert Cushman Murphy made his visit, in November 1912, the station at Grytviken had been in operation for eight years. In its first year it had taken a hundred whales, and each year the number had grown. The evidence of this success was everywhere to be seen:

> The whole shoreline of Cumberland Bay proved to be lined for miles with the bones of whales, mostly Humpbacks. Spinal columns, loose vertebrae, ribs, and jaws were piled in heaps and bulwarks along the waterline, and it was easy to count a hundred huge skulls within a stone's throw. The district is an enormous sepulcher of whales, yet no one can guess how many thousands of flensed carcasses have been

---

[24] The *Daisy* met an unfortunate end. During the First World War she was used as a merchant vessel, and was engaged in ferrying a cargo of beans to Europe when, on 29 October, 1916, she sprang a leak. The sea-water caused her cargo to swell, and the pressure of the expanding beans literally burst the brig asunder. She sank.

borne out to sea by the tide, and so have sunk their skeletons
in the deep.[25]

The station at Grytviken included the southernmost Post Office
in the world, and the southernmost Lutheran church, but it was
a very civilised outpost indeed. Larsen's residence was graced
with luxuriant palms and exotic flowering plants, which more
than dispelled the Antarctic gloom. Inside were a piano,
singing canaries, and more plants in a conservatory. Several
portraits, including one of King Haakon of Norway, looked
down from the walls, and Larsen lacked for little. On receiving
Murphy and Murphy's captain, he laid on an eight course
lunch, complete with butler, beer and "excellent . . . Havana
cigars". Murphy noted wryly that "it calls for a good constitu-
tion to endure this antarctic fare—otherwise a man could die of
gout".[26]

The men of Grytviken were certainly well looked after, with
provisions and comforts shipped from far off Europe. But they
were lonely and isolated. Later in his stay Murphy went
exploring with the station doctor and a man called Willberg,
the company secretary. The three skied, Murphy somewhat
inexpertly, up and down the valleys that cut into the hills
behind the station.

> Suddenly a distant whistle came to us through the still
> mountain air. Willberg and the doctor Look'd at each other
> in a wild surmise—but they were not silent on their peak.
> "*Harpon!*" they yelled with one voice, as they immediately
> began to make tracks for home, soon leaving me far behind.

The *Harpon* was a visiting ship that brought with her sixteen
bags of mail.

> Everything was excitement among the South Georgia popu-
> lace. All work apparently stopped, and while scores of men
> pored over letters from their loved ones ten thousand miles
> away, others were ripping open newspapers and chattering
> about an attempt on the life of Theodore Roosevelt, the
> election of Woodrow Wilson, a war that Greece and the
> other Balkan States are waging against Turkey, and about

precarious conditions in the Old World generally. With British, Russians, Germans, and other Europeans amid the larger group of Scandinavians here, the stage was all set for stirring discussions.

There were two letters for Murphy from his beloved Grace, and as he recorded in his log, "I know now that everything is well . . . All's right with the world!"[27]

Larsen offered Murphy the chance to go whaling on the *Fortuna*, the first steamer to hunt whales in the far south. It was a chance Murphy leaped at:

We were four hours steaming from Grytviken to the field in which we began to see whales in all directions. The commonest were Humpbacks and Finbacks, which can be very easily distinguished even at a great distance by their spouts, because the Finback has the tallest spout of any whale, whereas that of the Humpback . . . is short, bushy, and explosive. Humpbacks are preferred game. They are fat, easily approached, and small enough to handle with relatively little effort . . .

The captain and gunner of the *Fortuna* is a younger man than I, by name Lars Andersen. The jolly old Santa Claus of a mate is Johann Johansen. . . . I have seen practically nothing of seven or eight other members of the crew, most of whom are down in the bowels of the *Fortuna*, although one is now standing in a barrel at the masthead, with his eagle beak projecting just over the rim . . .

This steamer rolls so that all the while it is tossing an avalanche of water back and forth across its maindeck. Fortunately it was designed and built for semisubmarine operations. High combings keep the brine from pouring into the hold, and the cannon platform at the bow is reached by a catwalk well above the wet deck. Every now and then it still looks as though the prow and gunner were about to plunge into the arched trough of a wave, from which only an empty platform could rise again. Each time, however, the *Fortuna* thrusts up her snout, and remains in air. I have been spending most of my time on the bridge, where I could get wet only with spray.

[27] Murphy (1947) p 164.

The whale gun was loaded through the muzzle by three men, under the supervision of the mate, about two hours after we had left port. It is a short, thick cannon, balanced so delicately on hinges and a turnstile that the gunner can point it in almost any direction as easily as though it were a pistol. The mate rammed home the charge, which consisted of a muslin bag containing about a pound of large-grained black powder. Over this he placed a handful of hemp oakum, a rubber wad shaped like the bung of a cask, and then a mass of cotton fiber. All this pillowing is to prevent the harpoon being broken by the discharge. The harpoon weighs about one hundred pounds. Its shank fills the whole barrel of the cannon, and it is held in place by marline seizings, hitched around iron buttons on the muzzle. The part of the harpoon projecting from the gun has four hinged steel barbs, which are also lashed down tightly with marline. The lashings break when the harpooned whale tugs the line, and then the barbs open out like the ribs of an umbrella inside the body of the whale.

The last step in preparation is screwing the bomb point onto the tip of the harpoon. The bomb is a cast iron container of a charge of powder and a three-second time fuse. It is, of course, blown to fragments by explosion inside the whale, but the harpoon itself, which is forged from the finest Swedish steel, is usually undamaged or, at any rate, needs only a little subsequent straightening on the anvils of the smithy. It can thus be used over and over again, requiring only a new grenade for each discharge.

At one time during the morning, eleven steamers from Captain Larsen's company, or from those in other fiords, were within sight of the *Fortuna*, and the banging of harpoon guns was almost continuous. Humpbacks seem usually to travel in pairs. When we sighted the first two spouts, another steamer was heading toward these same animals, but we were closer and consequently beat our rival to their vicinity. After we had manoeuvered for some time, one came to the surface just off the port bow. Captain Andersen swung the cannon, the prelude to a great roar and a dense cloud of smoke. Then I heard the muffled crack of the bomb and, as soon as I could see anything, a whale, already dead, was slowly sinking, belly up, at the end of our heavy hemp foreganger. The latter is the part of the harpoon line that lies in a coil on a platform just under the gun. It is held in place with ties of marline, which part like cobwebs when the line goes out. Only the best of Italian hemp is sufficiently strong, supple, and elastic to follow the projectile without breaking. Forty fathoms of this hemp, representing a greater length than the range of the harpoon gun, are attached in turn to five hundred fathoms of Manila line. The latter rides over a snatch block connected with strong steel springs in the hold, and these prevent sudden and strong jerks from breaking the gear. The whole outfit represents a sort of gigantic counterpart of the rod, reel, and sensitive hands of a trout fisherman.

The winch quickly hoisted our sinking Humpback to the surface, after which a sharp pipe was jabbed into its body cavity and it was pumped full of air until it floated like a balloon. After the pipe was withdrawn, the hole was stopped with a wooden plug. The carcass was made fast by a chain ... The flukes were then sliced off with a cutting spade. This was because the whale's 'propeller' has a certain tort, one stiff fluke bending slightly downward and the other upward. It has been found that when whale carcasses are towed rapidly tailforemost, the tort of the flukes is sufficient to start the whole body spinning around its long axis, with momentum enough to break any chain or cable.

All these operations took only twelve or fifteen minutes, and I felt cheated because the kill had been so unexciting. We then resumed hunting, with our catch floating limberly along the port side of the *Fortuna*, and within half an hour Captain Andersen killed a second Humpback. The process looked more than ever like murder with ease and no trace of uncertainty, but the officers assured me that it is not always so simple. They say that two harpoons are frequently needed and that three, four, or even five are not highly exceptional.[28]

The officers explained to Murphy that the bomb must blow up in the chest cavity of the whale if it is to cause practically instantaneous death. The back end of the whale contains large masses of muscle and other tissues that are not nearly so easily shattered as the lungs, and this hind portion of the whale may be the only target if the whale has finished spouting and

[28] Murphy (1947) pp 147–9.

inhaling and has begun to submerge. Besides, the effort was not always "with no trace of uncertainty". Later in the season a whaling cannon broke its mountings as the weapon was fired, instantly killing a gunner named Beckmann. Beckmann was captain of the *Don Ernesto*, the newest whale hunting ship down south. He was twenty-seven when he died, and left a wife and four children back in Norway.

Murphy may have felt cheated, but like a good naturalist he kept close watch on the proceedings:

> I noted that a neat little white notch was cut in the front edge of a fluke stump on each of our chained carcasses. The purpose of this was to inform the flensers at the shore station that they were to recover only a single harpoon from the body of each victim. The first thing that the men with the long-handled knives look for is the harpoon count, as shown in this way.[29]

In any case, the disappointment was about to be mitigated:

> . . . [T]he man in the barrel uttered some guttural croaks which were translated for me as "Blue whale, dead ahead!"
>
> We were all agog to outguess this leviathan, which was engulfing krill with the joy of Eden, unconscious of a foe armed with devices worse than those of the serpent. Whispered messages passed down the speaking tube.
>
> "Half speed."
>
> The gunner ran forward to his platform, where the harpoon was then being rammed into the weapon and seized in place. Its bomb point began to sway back and forth over water that would soon run red. The whale slick was closer. Engines at quarter speed—it might breach within range at any instant.
>
> "Under the bow!"
>
> Up swung the butt of the swivel gun. A flash and deafening detonation split the frosty air, and this time I actually saw the iron, flinging its wild tail of line, crash into the gray hulk. Three seconds later—a long while it seemed—came the strange, faraway crack of the bomb. Had it reached his lungs?
>
> No, the shot was too far aft for an instant kill. The

[29] Murphy (1921) pp 149–50.

wounded whale summoned his final energy and towed us, against the drag of the winch, against the whirl of the propellers that had responded to full speed astern.

No being can reveal more marvelous grace than a whale. Do not think of them as shapeless, as I once did, because of seeing only bloated carcasses washed ashore on Long Island beaches, with all the firmness and streamline of the body gone. Envision rather, this magnificent Blue whale, as shapely as a mackerel, spending his last ounce of strength and life in a hopeless contest against cool, unmoved, insensate man. Sheer beauty, symmetry, utter perfection of form and movement, were more impressive than even the whale's incomparable bulk, which dwarfed the hull of the *Fortuna*.[30]

So much for the hunt: two Humpbacks and a Blue in a matter of hours. Then came the business of getting them back to the land station at Grytviken. A three or four hour journey unencumbered to the whaling grounds was a very different matter from the reverse, towing three huge carcasses. At half, or quarter, speed, with a fog enfolded around them, it took all afternoon and most of the night to get back to Grytviken. And that day's hunting had not been particularly good; a week later Murphy saw the *Fortuna* puffing into port:

> . . . towing nine dead whales. Except for her superstructure, she was practically concealed by the bloated carcasses . . . She had killed also a tenth whale, but being unable to find room to make it fast, had left it . . . afloat to be picked up by one of the company's other steamers! . . . A steamer preceding the *Fortuna* brought in six whales this morning, and the water off the flensing slip is now covered with a huge raft of carcasses.[31]

The *Fortuna* reached port at about one o'clock in the morning. After turning in for the rest of the night, Murphy went ashore next day to watch the flensing:

> Men carrying sharp curved knives, which have handles four or five feet long, make longitudinal slits in the whales, and the blubber is then peeled off from end to end by steel cables

[30] Murphy (1947) p 150.
[31] Murphy (1947) pp 159–60.

wound on the drums of winches. Nearly all the men are
young, strapping fellows, strong and graceful. They wear
suits of soft thin leather, and boots with spiked soles that
enable them to walk all over a whale carcass . . .

Day before yesterday sixty whales were received at
Grytviken slip. A pod of carcasses lay afloat, in addition to
those lying side by side on the incline. Now they are all gone,
except for two or three still in the works; the rest are already
oil and fertilizer. Last year[32] 51,000 fifty-gallon barrels of
whale oil were made at this plant alone, and the prospects of
the current season, which will end next March, are still
better. Since six additional whaling stations are already
operating in other fiords of South Georgia, you can imagine
what the slaughter amounts to.[33]

The slaughter was immensely worthwhile. Murphy faithfully
reported what he heard:

The gossip at Grytviken is that Captain Larsen's company
has paid annual dividends to the stockholders of from 40 to
130 per cent ever since the station was founded. I don't know
what the machinery cost, but it is said that it was transported
and erected here . . . for only $10,000. Twenty-five thousand
dollars bought the first chaser with which the company
started, and at the end of the six months' summer season the
first shipment of oil was sent to market. The first year's
dividends were regarded as low, being only about 40 per cent
on the invested capital, but at the end of the second year the
stockholders received a dividend of 100 per cent.[34]

The British, who administered the island bases, became
slightly wary of depleting the whales. In 1906, the government
issued an order regulating the whale fishery of the Falkland
Islands and neighbouring waters. There was a £25 permit to
fish, and a royalty on each whale. Right whales brought in ten
pounds, Sperm whales ten shillings, and other whales five
shillings. That was repealed two years later, although licences
were still payable. Mothers and calves were not to be taken. All
parts of the whale were to be used, not just the valuable

---

[32] 1911
[33] Murphy (1947) p 153.
[34] Murphy (1947) p 161. The gossip was absolutely correct, as Tønnessen and
Johnsen (1982) document in such detail.

blubber, and a tax was levied on every barrel of whale oil processed in the islands.

These measures were no doubt well intentioned, but they had the unfortunate effect of hastening the end of Antarctic whaling. The Norwegians deeply resented the British controls, and set about finding a way to escape from their physical need for the British-controlled island stations. The solution was the floating factory ship. There were already processing plants on pontoons in the waters of Cumberland Bay, and a primitive factory on floats put to sea in 1914. Ten years later the modern factory ship appeared. The *Lancing* was the first, a huge vessel fed by a fleet of smaller catcher boats. The *Lancing* was equipped with a slipway up which the carcasses brought to her by the catcher boats could be winched. On the decks the whales were flensed and the blubber rendered down in vast pots. The meat was hardly used. With cannons and powered boats, whaling could no longer be thought of as hunting in the noble sense; it was slaughter.

The whales, by their behaviour, in a sense contributed to their own downfall, although of course they could not help but do so. They came to the Antarctic to feed, and therefore could be found congregating around local abundances of plankton. Furthermore, they had to surface to breathe, and each time they surfaced they were vulnerable to the catcher boats' harpoons. Rather than fleeing when one of their number was hit, they often milled around trying to help the wounded animal, making the whalers' task that much easier.

The factory boats made huge inroads. Between 1910 and 1940 the catch rose from 12,000 to 40,000 whales a year, with a peak in 1931 of 55,000 animals, 5 million tonnes of whale. Norway and the UK had the Antarctic to themselves until 1934, when the Japanese arrived. The Germans joined them the following year. After the Second World War the USSR, USA, Australia and others joined the scrum. By the middle of the 1960s, the ancient pattern had repeated itself, and most of the countries no longer found it worthwhile travelling down to the Antarctic for the few whales that remained. Only Japan and the Soviet Union continued.

Whales were more valuable to the Japanese than to any other nation, because the Japanese had a well-developed market for

whale meat. Other countries used only the oil. As a result Japan
found it profitable to continue whaling after other nations had
abandoned the Antarctic, and built up its fleet at good prices by
buying whaling vessels from Norway, the UK, and the Netherlands. They are still there today, pursuing the smallest of the
rorquals, the Minke whale. Before, when there were great
whales left to hunt, there was never any virtue in capturing
Minke whales. Now they support the remnants of the industry,
not only in Antarctica but also off the northern coast of Norway.
From a peak of 29,000 animals, the annual catch has dropped to
about 4,000, and even that is scheduled to end.

The history of whaling reveals one inescapable fact: in every
case, commercial whaling has caused a severe decline in the
stocks. The aboriginal people who took whales did so to fulfil
their own needs. There was little question of accumulating a
surplus for trade, and no incentive to risk all for another whale.
Even with their slow rate of reproduction the whales could
sustain this harvest. Commercial whalers, however, saw whales
as a source of wealth. Each one taken represented more money,
and there was no incentive to be prudent. Technology kept

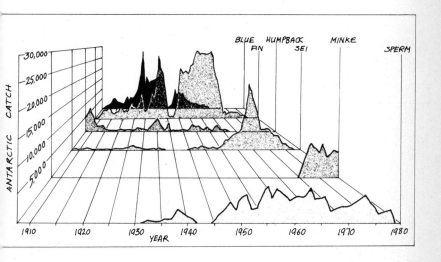

coming to the rescue of the whalers. As the coastal Right whales declined, better ocean-going vessels made it possible to sail further afield and exploit whale stocks elsewhere. As Right and Sperm whales elsewhere declined, Svend Foyn's harpoon gun allowed the whalers to go for the larger, swifter whales—the Blues, Fins, Seis and Humpbacks—that had eluded them before. When the northern populations were exhausted the whalers went down south to the Antarctic, and there they repeated the whole sorry history of whaling.

Between 1925 and 1938 they took Blue whales. The effort in catching a Blue was not much greater than that required for a Fin or a Sei and the yield of oil was considerably higher. By 1931 Blues were becoming scarcer, so the whalers turned their attention to Fin whales, roughly half the size of Blue whales. As the take of Blues decreased, that of Fins increased. In 1938, for the first time, Antarctic whalers took more Fin whales than Blues. The Fin whale predominated until about 1960, when it too became harder to find. The Sei whale took over, forming the bulk of the catch between 1965 and 1973. Then it was the turn of the relatively small Minke whale, which until that date had been dismissed by whalers as not worth the effort of catching. With Blues, Fins and Seis gone, the Minkes became the mainstay of the Antarctic whalers. Now, for the first time, there is a chance that the whalers will stop their harvest of Minkes before they simply run out of Minkes to take. The history of regulation, however, does not make me optimistic.

Control

But their wild exultation was suddenly checked
     When the jailer informed them, with tears,
Such a sentence would have not the slightest effect,
     As the pig had been dead for some years.

Fit the Sixth: The Barrister's Dream

The great whales belong to nobody, and to everybody. In the struggle to exploit them the spoils go to the stronger and the swifter. Whichever nation puts most effort into whaling will get most of the resources. Modern whaling in the Antarctic could, therefore, have been subjected to overzealous competition from the very start, but there were two factors that went some way to preventing this: the first was that very few nations were in fact exploiting the Antarctic stocks, only Norway and the United Kingdom until Japan joined them in 1934; the second was the high cost of entry into whaling. Special ships had to be built and manned by skilled people, barriers that prevented new competitors joining the scramble. The enormous profits to be had from whaling, coupled with the obvious losses that would occur if competition were unbridled, acted with these two predisposing factors to make some co-operation a valid alternative to competition.

The co-operation between the United Kingdom and Norway was not the first attempt to control whaling. There had been quite a few earlier laws and agreements, which took three distinct forms. There was national legislation, enacted by one country to prevent others taking advantage of whales in its waters. The Russians, for example, banned foreign whalers from the Bering Sea in 1821. This was one of the first pieces of national protective legislation, although it was designed to protect Russian whalers rather than whales. Norway passed laws in 1903, and again in 1929, regulating the industry at home and abroad and making sure that Norwegians were not allowed to sell their expertise to other whaling countries. The British government introduced its own restrictions, aimed particularly at Norway, to control whaling around the British Isles and in the Antarctic dependencies, and there were several other pieces of national legislation.

The second form of control was the private agreement within the industry, which began with the great whale-oil flood of 1931. British and Norwegian boats caught so many Blue whales in the Antarctic that season that the bottom dropped out of the market. To prevent a recurrence, the members of the Association of Whaling Companies reached a deal which limited their total catch the following season to 320,000 tonnes, divided among the sixteen expeditions that were party to the agreement. This agreement set a precedent for many subsequent decisions; it emerged after the crisis had happened. Furthermore, it enshrined a principle that returned to make nonsense of later attempts at control. Instead of setting limits for each species, and instead of setting them in terms of the weight of the animals, it adopted a notional Blue whale unit. The oil yield of one Blue whale was reckoned to be equal to two Fin whales, two and a half Humpbacks, or six Sei whales. The Blue whale unit allowed whalers to take the most profitable whales they came across regardless of how rare that particular species might be. The whaler might not encounter many Blue whales, because there were already very few left, but when he did find one he could kill it. Not for another forty years was this biological nonsense abandoned. The agreement also made no allowance for whale meat, which at that time was the least important part of the whale. Indeed, it was generally tipped overboard.

The Association of Whaling Companies' agreement lasted for just two seasons, during which time Japan and Germany joined the Antarctic whalers. In 1935 the third form of control began when the first international agreement came into force. This protected all Right whales completely. (Blanket protection was extended to the Gray whale a couple of years later.) It also prohibited the take of sexually immature animals and nursing females and their calves. The Second World War interrupted whaling but in 1944 the Allied signatories of the 1935 agreement met in London to decide how to proceed when the war was over. They agreed to limit the catch to 16,000 Blue whale units, roughly two-thirds of the average annual catch before the war, and specified that the Antarctic season should last no longer than four months.

The war itself had hit whaling fleets hard. Ships, both factories and catchers, were commandeered for wartime uses, even though they were suitable for little other than whaling. By the time peace was restored fewer than a quarter of the forty-one factory ships at work in 1939 remained functional. But in addition to destroying the means of supply, war had also created enormous demand for edible oils and fats. The whaling fleet was hastily rebuilt, and by 1948 Britain and Norway had commissioned seven new factory ships to process the whales of the Antarctic. The cost of a whaling fleet—the factory and its catchers—was about $10 million just after the war, but the demand for whale oil made it worthwhile, and many countries were eager to get involved.

There might have been appalling conflict, but in 1946 all fourteen nations involved in whaling on the high seas met in Washington to negotiate an International Convention for the Regulation of Whaling. This convention established the International Whaling Commission and a very important principle. The catch quotas and other matters of control were devolved to a separate document, the Schedule of the convention. The members of the IWC could modify this schedule directly, with no need for prolonged diplomatic wrangling, although it needed a three-quarters majority of those voting to amend the schedule. This provision is stronger than the two-thirds majority more common in international treaties, and was intended to strengthen the convention. It does so, but it also builds considerable inertia into the system; once on a particular track the IWC is very hard to deflect.

Counterbalancing the potentially beneficial provision of a separate, and thus malleable, schedule were two decisions with less welcome consequences. The IWC, following the Association of Whaling Companies, set its limits in Blue whale units. It also made no attempt to divide the total quota between the different whaling concerns, which was seen as a matter for private agreement. Both of these principles led to the same predictable result: a mad scramble for the most profitable Blue whales, regardless of their depletion relative to other whales.

The structure of the IWC, and the number of member nations, has fluctuated over the years, but in essence there are three important tiers. The Scientific Committee and the Technical Committee are composed of those member nations that wish to serve on them. The Scientific Committee assesses the state of the whales and makes recommendations to the Technical Committee. The Technical Committee forwards these, and other matters, to the full plenary sessions of the IWC. It is here that a three-quarters majority can change the schedule.[1]

(Most important of all, it could be argued, is an organisational tier that does not really belong to the IWC at all. Hanging about on the fringes of the IWC meetings are the so-called NGOs, non-governmental observers, or organisations. These are lobbyists from all points of the whaling spectrum, and all levels of integrity and competence. Anyone with an office in at least three countries—from the Assembly of Rabbis to Greenpeace International to the All Japan Seamen's Union—can become an NGO. They provide a very effective service in keeping commissioners informed of the issues, important and otherwise. And since 1978, when journalists were excluded from the IWC's deliberations after a conservationist masquerading as a journalist attacked the Japanese delegation with blood-red paint, they have also kept the press fed with stories.)

The purpose of the convention is explicitly "to provide for the proper conservation of whale stocks" and "thus make possible the orderly development of the whaling industry". The implication is that the whale stocks come before the whaling industry. The convention further states that important decisions "shall be based on scientific findings". By and large these fine principles were, in the early days of the IWC at least, honoured far more in the breach.

For example, the schedule as drawn up in 1946 created sanctuary areas in the Antarctic and expressly forbade the capture of Humpback whales anywhere in the Antarctic because stocks were so low. At its very first meeting, held in London in May 1949, the IWC established a Scientific Subcommittee to guide its decisions. It went on to lift the total ban on killing Humpbacks and to open the sanctuary. This set the tone for years to come.

---

[1] Each of the committees can also set up sub-committees to report back on specific problems, a useful delaying tactic.

As a gross oversimplification, I believe it is fair to say that the Scientific Sub-committee did its best to warn the IWC that the stocks were being hit very hard. The scientists kept recommending caution—lower quotas, shorter seasons, sanctuary areas—but the IWC continued to give the whalers—who, after all, *were* the IWC at that time—what they demanded. And when the whalers did not get what they wanted they either quit the IWC or else objected, quite legitimately under the rules, to its decisions.

Objections, enshrined in Article V of the convention, are a peculiar feature of the IWC. All members have the right to object to any decision of the IWC, provided they do so within ninety days of the meeting. An objection exempts that nation from that decision, and extends the ninety-day period to 180 days for other countries that might want to object. This provision of the convention effectively robs the IWC of any power to enforce its decisions. It has in any case no jurisdiction over whaling by countries that choose not to belong to it, and a member country that objects under Article V of the convention is, quite simply, not bound by it.

There is a fine irony about the whole business of objections. At the 1946 meetings that led to the convention, New Zealand, Norway and the United Kingdom were strongly opposed to the idea of permitting escape from IWC decisions. Article V made it into the convention on the insistence of the United States. Now, forty years on, the only effective sanctions to implement decisions of the IWC are provided by the US.

America pulled the IWC's teeth in the first place, and America now provides the IWC's only dentition. Two amendments to US fisheries laws enable that government to prevent access to fish in US waters and to block imports of fisheries products from any country that "diminishes the effectiveness" of the IWC. The first is called the Pelly amendment, and if a country is certified under its terms the President has the option of banning any or all fisheries imports from that country. The second—which blocks access to US waters—is called the Packwood/Magnuson amendment, and is supposed to be mandatory. These are big threats to miscreants, and should have been able to bring whaling to a dead stop. In reality, however, things are different. (See "Sleazy Deals", p 171.) For example, the Commerce Department has so far declined to sanction

Japan for whaling in contravention of several IWC decisions, although the USSR has been penalised for similar breaches. Conservationists obtained a legal ruling that the Secretary of Commerce was bound to invoke the provisions of the amendments, and courts decided in favour of the conservationists three times. The government appealed one last time, and the case found itself in the Supreme Court. On 30 June, 1986, by the narrowest of margins (5–4), that court reversed the earlier decisions. The Court ruled that the executive branch has special discretion when it comes to applying laws that might have an impact on foreign relations. The US is under no compulsion to use its teeth to bring whalers to order.

It has been said—often—that the IWC was a whalers' club, which watched over the destruction of the whales while its members made enormous profits. A detailed look at some episodes from its history makes it hard to avoid that impression.

In 1959, for example, there was a crisis. Before the annual meeting the five nations with fleets in the Antarctic—Japan, the Netherlands, Norway, the UK and the USSR—met to allocate national quotas. They could not agree, and Norway and the Netherlands left the IWC. (Japan gave notice to quit, but did not in fact do so.) The IWC responded by lengthening the season and opening it ten days earlier, despite the scientists' fears that this would result in a greater number of pregnant females being killed. The Antarctic sanctuary was left open to whalers. The quota was set at 15,000 Blue whale units, but although the catch itself was higher than in the previous seven seasons the whales themselves were much smaller. More than half of the Blue whales taken were immature animals. This is a classic sign of biological overexploitation. It means that the population has too few adults, and when hunters start taking too many immature animals they are removing the very capital on which any sustained exploitation depends.

The Chairman's Report of that meeting is chilling, considering that the IWC's convention requires it to "protect all species of whales from further overfishing". He said:

> Conscious of the importance of maintaining the Convention, the Commission showed a willingness to consider making some increase in the Antarctic catch if thereby the loss of

114

those member countries which had given notice of withdrawal could be averted.[2]

The following year, 1960, the IWC established another subcommittee of three scientists, to be known as the Committee of Three Scientists, that would investigate stocks and make recommendations. (The three were K. R. Allen, Doug Chapman and Sidney Holt; the three became the four when John Gulland joined them the following year.) The commission also decided to set no quota at all for the Antarctic, probably in an attempt to woo Norway and the Netherlands back into the fold. The catch that year was 16,433 Blue whale units, the highest in the history of the IWC and never to be repeated, even though there were no restrictions in the following season either. Of the 1,750 Blue whales caught, no fewer than 1,237 (71 per cent) were immature.

Three years later, in 1963, the IWC had the reports of three separate committees of scientists before it. None of their recommendations was new. The Blue whale unit should be abolished to enable species-by-species quotas that would protect those whales that needed it. There should be a complete ban on the taking of Blues and Humpbacks in the Antarctic. And the quota should be much lower. In response to these clear and unequivocal suggestions, the IWC voted some limited protection for the Blues and Humpbacks, but retained the Blue whale unit.

In 1964 the Japanese fleet, by now accounting for considerably more than half the effort in the Antarctic with 100 catchers and seven factories, caught not a single Blue whale.

In 1965 the IWC eventually banned the taking of Blue whales in the Antarctic.

The IWC finally abandoned the Blue whale unit in 1972.

The IWC had established a tradition of doing too little too late. It asked scientists to make recommendations and then found reasons not to act upon those recommendations. Towards the end of the 1960s this state of affairs had become extremely irksome to the scientists involved. According to Sidney Holt, the situation was becoming so bad that control of the IWC had

[2] Friends of the Earth (1978) p 25.

to be wrested from the whalers. The method chosen was a resolution put before the 1972 United Nations Conference on the Human Environment in Stockholm. This called for a ten-year moratorium in commercial whaling, and was passed by fifty-three votes to nought (with twelve abstentions).

The United States and the United Kingdom attempted to get a similar resolution through the IWC that year, but without success because there was no time to get their proposal onto the agenda. It failed by four votes to six (with four abstentions) in 1973. Similar attempts in 1974 and 1975 also failed, but in 1974 the IWC did pass a resolution that changed its way of approaching quotas. This resolution, which cropped up in the form of a combined Australian/Danish amendment to the moratorium, formed the basis of the so-called New Management Procedure. The New Management Procedure emphatically stated that the IWC was to take a long-term view of the whaling industry, so that a pause in commercial whaling could, if the evidence warranted it, be implemented because it was in the best long-term interests of whalers. It also enshrined three principles of management: whale stocks should be held at an optimal level to allow the maximum sustainable yield to be removed each year; any stock below that optimal level was to receive protected status, with the intention that the stock recover to optimal level as quickly as possible; and each stock of every species was to be treated separately, catch limits being agreed every year. (See "Science" for more details, p 131.)

These are not bad guidelines for wildlife management. The notion of maximum sustainable yield is a familiar one to ecologists. Every animal population, when it is in balance with its environment, is relatively stable. Births are roughly equal to deaths. But there is a surplus of untapped capacity. The population is always capable of reproducing more quickly than it actually does. This surplus can be exploited by reducing the population to some optimal level, at which the number of births is at a maximum. There is one problem, however, especially when this principle is applied to whales: it requires an accurate knowledge of the population dynamics of the species concerned. One needs to know how many whales there are, how many there were in the original pristine population, how many die of natural causes, how many are born each year, and many other parameters. For most whales this information is not very reliable. For example, even the best estimates of the size of a whale population are just that—estimates. Scientists obtain the final figure by extrapolation from a sample, and their conclusions tend to have at least a twofold area of uncertainty: the numbers may be half the estimate, or twice the estimate. And estimates of the rate of reproduction vary even more; some whale biologists think it is as high as 10 per cent (that is 10 whales born each year for every hundred adults), while others think it may be below 2 per cent. So even though the mathematical equations do in the end offer a precise figure for maximum sustainable yield, the range of variability in the estimates of population dynamics implies an eightfold range in the final figure of maximum sustainable yield. If one were totally wrong, one might be taking eight times more whales than one "should" be.

There were other problems with the New Management Procedure too, to do with the collection and analysis of data. Those species that had long been exploited—the Blues, Humpbacks, Fins and Seis—were quickly shown to be far below optimal level and could be given protection status. But many other exploitative endeavours, for example the newly burgeoning Minke and Bryde's whale hunts, and the long-established Fin and Sperm whale hunts in the North Atlantic, could not be assessed. There simply was not the information on which to base New Management Procedure calculations, in the case of the North Atlantic because the whalers refused to gather and analyse the data. As a result these hunts tended to receive quotas that were simply an average of past performances by the whalers, a recipe for disaster if too many were already being taken. The New Management Procedure had failed to provide an "indeterminate" category for scientists to use when they were genuinely unsure of the status of a stock, and as a result those stocks that might have been protected before they became severely depleted were not.

So although the intentions of the New Management Procedure were good, they were very hard to implement effectively. The whalers could always use the higher estimates, while the conservationists used the lower ones, and in truth there was no easy way to settle the issue. It became a question of who should get the benefit of the doubt, the whales or the whalers.

The balance began to shift, ever so slowly, as the New Management Procedure became accepted. The years immediately after 1974 were uneventful, but by 1979 a new spirit seemed to emerge, guided by the Whaling Convention's statement that whales were a resource for all the people of the world. If the IWC had been a whalers' club, that was not because of any rules in the convention. Any country, even land-locked states with no commercial interest in whales whatsoever, was free to join. All they had to do was register with the IWC and pay their money. Conservationists began to lobby sympathetic countries and draft them into the IWC fold. The money for those countries' IWC fees, and support for their delegations, often came from private funds as politically astute conservationists sought to change the IWC from within. The goal was to create a three-quarter majority of anti-whaling nations.

It was in 1979 that the newly independent Indian Ocean state of Seychelles joined the commission. The Seychelles was very ably represented by Lyall Watson and Sidney Holt, who between them displayed a joint mastery of diplomatic and scientific techniques.

The year before, Australia had made a most courageous decision, that all whaling was indefensible. This was the conclusion of a Royal Commission conducted by Sir Sydney Frost, a highly respected judge. Frost's report stated that Australia should close the single coastal whaling station still operating and press for an immediate and permanent end to all whaling, not just a temporary moratorium on commercial catches. This was an ethical position, not subject to argument, and it is to Australia's credit that she has stuck by that decision through all the vicissitudes of the past years.

While quotas were coming down as the New Management Procedure was being applied, the whalers within the IWC were not idle. They set about securing the whales denied to them by the IWC's quotas, and they did so by the simplest expedient: buying whale products from countries that were not members of the IWC. Brazil, Chile, Peru, South Korea and Spain all benefited from massive Japanese investment. A report on the subversion of the IWC describes them as "Japanese whaling colonies", with "Japanese capital, vessels, equipment and expertise, as well as import markets for whale meat".[3] Whaling imperialism was one response to reduced quotas. The

[3] Carter and Thornton (1985) p 4.

118

infamous pirate whaling was another, and the two were changed for all time by the exposure conservationists gave them at the 1979 meeting.

The pirate whalers flew under flags of convenience—Cyprus, Somalia, even Liechtenstein—and operated outside the IWC. Their ships roamed the seas taking any whales they could, including Blues and Humpbacks. It was Japan and Norway who brought an early example of pirate whaling to the IWC's attention, at the 1955 annual meeting in Moscow. The whaling fleet owned and operated by Aristotle Onassis roamed outside the IWC between 1950 and 1955, causing untold damage to whale stocks and creating tens of millions of dollars of profit. The whalers barred Onassis from their club: "On quitting [pirate whaling] Onassis observed he had been overwhelmed by more monopolistic IWC interests whose motives were no different to his own."[4] This was shown to be entirely accurate in subsequent years by the much more widely known case of the pirate whaler *Sierra*, owned by Norwegians, operated by South Africans, under Japanese supervision, selling oil to Norway and the EEC and meat to Japan. The *Sierra* was just one of the pirate whalers whose intricate and well-concealed lifelines to legitimate whaling conservationists laid bare at the 1979 meeting.

The revelations about the pirate whalers caused quite a stir, with important consequences. Under threat of sanctions from the US, Japan announced that it had passed laws forbidding the import of whale products from any country that was not a member of the IWC. The wording allowed whale meat caught by non-IWC whalers to be laundered through an IWC member, a loophole that Japan exploited for at least two further seasons, until those connections too were laid bare. Japan's new legitimacy also caused a setback for conservation, as South Korea, Peru and Spain joined the IWC that year, to be followed in subsequent years by other countries that had been whaling on behalf of the Japanese. One vote for whaling, because of the structure of the IWC, was equivalent to three conservationists.

Conservationists outside the IWC grew weary, but were heartened by two bits of news. One of the pirate whalers turned

[4] Carter and Thornton (1985) p 4.

turtle and sank while hauling aboard just one more Fin whale. And the *Sierra* herself was rammed by militant conservationists in the *Sea Shepherd*. Other pirates too were neutralised, either by direct action or by publicity, and although there are still occasional reports of whalers operating where none should be, these are fewer and the defeat of the pirates could be considered one of the great victories of the conservationists. They were successful within the IWC too.

The United States once again proposed a temporary moratorium on commercial whaling. Jean-Paul Fortom-Gouin, a wealthy Frenchman, was a conservationist gadfly like Watson. He had become the driving force behind the delegation from Panama, and amended the American proposal on behalf of his delegation. Fortom-Gouin suggested splitting the ban into coastal and pelagic portions; if the IWC passed both it would be equivalent to the entire US motion. After much discussion another amendment appeared. The Danish commissioner amended the Panamanian amendment to exclude Minke whales—the only species then being caught in any numbers by the pelagic factory fleets. The ban on factory fleets sailed through, while the coastal ban sank, which it surely was intended to do. The main effect of the 1979 meeting was thus to prevent the Soviet Union using its factory fleet to catch Sperm whales.

The Seychelles had come to the meeting with a proposal for a gigantic sanctuary for all whales in the Indian Ocean. This had to be trimmed to allow the Japanese and Soviets to continue taking Minkes in the Antarctic, but it too was passed in its modified form. That also was a gain for the whales.

The haggling over the quotas revealed the price that had been paid for these gains. In almost every single instance the quota agreed was at the high end of the scientific estimates. In some cases, notably that of the Bowhead whales of the Arctic (see "Bowheads", p 151) it went in direct opposition to the scientists. The thirty-second meeting, in 1980, was especially notable as the year of the Bowhead. The US managed to secure a three-year quota for Bowheads, which at least would leave it free to pursue conservation rather than quotas in the following two years, but at what cost? Again there was a proposal on the table for a complete moratorium, but it failed, partly because the Scientific Committee had not given the moratorium its unanimous backing. This may have been because a blanket ban was seen as a return to the days of the Blue whale unit, when whale stocks were lumped together for assessment.

In the course of the 1980 debate on the moratorium the delegate from the USSR, I. V. Nikoronov, made an astonishing statement outlining his country's perception of the IWC. He said:

> One group of countries strives to use marine resources in a rational way and these countries carry out investigations and obtain information which forms the basis of the activity of the scientific committee. The other group takes a contrary stand. Actually without making any practical contribution to investigations this group strives for a ban on whaling and leads the Commission to a dead end.

That needs some decoding. The first group of countries is the whaling nations. Their "investigations" consist of hunting whales, their "information" the bodies of those whales. The "other group" is the countries that feel whales are being overexploited; their lack of "practical contribution" is their failure to kill whales.

The scientists I spoke to afterwards were livid. Quite apart from what they saw as Nikoronov's misleading usage of words like "rational", there was the fact that most of the scientists who had made any contribution to a better understanding of whale demographics were from non-whaling countries. They resented the suggestion that only dead whales provided useful information, not only because it is not true but even more so because the scientists had for years tried in vain to get the whalers, especially the USSR, to release up-to-date data on the catches. Without those data the scientific conclusions were indeed less reliable than they might have been.

The Soviet justification for a continued take was even more astonishing:

> At present whaling has been reduced to such a low level that the removals from the stocks have become minimal with no risk of over-exploitation. Everybody appears to understand this. However, representatives of some countries, guided by subjective reasons disguised as noble aims of the conservation of living resources, attempt to undermine the principles of the existing whaling convention.

121

In other words, because the quotas were already so low, no further harm could be done to the whales, and never mind the conditions that had made such low quotas necessary.

The IWC also had to consider a move to protect Sperm whales, still being taken by coastal whalers, and illegally by the factory fleet of the USSR. It would essentially complete the previous year's ban on taking Sperm whales by factory fleets. It failed by fourteen votes to six, one vote adrift. Had Canada, a non-whaling nation, voted yes rather than no the Scientific Committee's recommendation of a Sperm whale ban would have been accepted. Mac Mercer, the Canadian commissioner, was accompanied everywhere—even to the toilet—by burly security guards, and the IWC did not deny that threats on his life had been made. But nobody was able to discover why he had voted the way he did. To this day, I do not know why Canada supported the USSR on this issue.

The whalers now had an immovable bloc of nine nations —Brazil, Chile, Iceland, Japan, Korea, Norway, Peru, Spain, and the USSR—which proved more than a match for the rest when it came to quotas that year. Although they perhaps did not get as many whales as they wanted, they generally got more than the Scientific Committee had been willing to allow.

The conservationists returned to the fore the following year, 1981. China, India, Jamaica and St Lucia joined the IWC before the annual meeting started, and St Vincent, Costa Rica and Uruguay all produced conservation-minded commissioners before the decisive votes. It was the year of the "non-moratorium" on Sperm whales.

The first crack in the whaling bloc appeared as Chile, Norway and Spain abstained from a crucial vote on the Sperm whale ban when it was discussed in Technical Committee. That was on the Tuesday night. The vote did not come up again until the final plenary session on Saturday, and one can only guess at the bargains that were struck in the interim. I saw the Japanese and Spanish delegation dining together in Brighton's most expensive fish restaurant, an appropriate spot for wheelers and dealers. I saw Lyall Watson locked in conversation with Kunio Yonezawa, the Japanese commissioner. Rumours flew, but in the event the Sperm whale moratorium was achieved in an astonishingly clever way.

In the past, and again in 1981, calls for a moratorium had always been phrased in exactly those terms, and it was an easy matter for the Japanese, and like-minded nations, to object to them on the grounds that there was no scientific evidence that favoured a moratorium. This is undeniably true. The scientific evidence indicates, more than anything else, uncertainty; it always has. The whalers want to have the benefit of this uncertainty for themselves, while the conservationists want to give it to the whales. This time, however, the Seychelles came up with a nifty alternative. Instead of calling for a moratorium, Lyall Watson asked simply that "catch limits . . . be set at zero". Netherlands, France and the UK dropped their own proposals and joined the Seychelles, and the Technical Committee agreed that this was "an appropriate response to the depletion and uncertainty of assessment".

The only Sperm whales being taken at the time were 300 off the west coast of South America, which had already been set to be abolished the following year, 130 in the North Atlantic, and 890 in the coastal waters around Japan. Catch limits on all stocks would be set at zero, except for Japan's, which would be "unclassified" for the time being, and all catch limits could— given the necessary majority—be open to revision by the commission at any time. A footnote prohibited the Japanese from taking any Sperm whales until their stock had been reassessed and classified and a quota awarded.

This form of words overcame almost all objections that could be mustered. The provision for reassessment at any time gave Kunio Yonezawa, head of the Japanese delegation, the chance to produce good scientific evidence to resume whaling some time in the future. "Our concern was to give Yonezawa something he could live with," I was told with surprising directness by Cornelia Durrant, one of the Seychelles' delegation.

The Jamaican commissioner, who throughout the proceedings displayed a fine sense of idealism coupled with long parliamentary experience, supported the proposal but exposed it for what it was. "The whole thing could have been said in one straight line, instead of a number of ellipses," he said, though personally I doubt that it could. And Richard Packer, the United Kingdom's commissioner, confessed to being "somewhat disappointed" by the form of words. He nevertheless felt able to commend the proposal to the commission as "the best outcome that can be achieved, as my American colleagues would say, at this moment in time, or in other words, now". It was late in the day, early on Saturday morning.

Japan naturally "had a few difficulties with the principles involved. It has been a consistent policy of my delegation," Yonezawa said, "not to vote yes to those proposals where we cannot find scientific basis. Moratorium is one such proposal, and I regret that my delegation cannot associate with this proposal." It was a gracious statement. In the final count only Japan voted against the *de facto* moratorium; Iceland and the USSR (and China, though that does not count as China had yet to vote decisively on any proposal in this, its first meeting of the IWC) abstained.

As a result of this neat piece of ju-jitsu the weight of the commission had successfully been turned in favour of the Sperm whales. Just as in the past the whalers could generally muster sufficient support to block change, the conservationists would now be able to resist attempts to open the Sperm whale hunt. (In fact, the vote did nothing to stop the Japanese taking Sperm whales; see "Sleazy Deals", p 171.)

There were other decisions at the 1981 meeting, of course. Quotas were set for other whales, often in excess of prudent scientific advice. And there was a total ban on the so-called cold-grenade harpoon, a non-explosive harpoon which does not damage as much meat in a small Minke whale but probably takes longer to kill the beast than an explosive harpoon would.

The crucial year, for conservationists, was 1982. That was the year in which the International Whaling Commission decided, by twenty-five votes to seven (with five abstentions) to phase out commercial whaling. (The UN, recall, had made a similar decision in Stockholm a decade earlier.) The start was to have been in 1984, but was delayed for three years rather than two, to the 1985/86 Antarctic season, and the moratorium was scheduled to be reassessed by 1990 at the latest. But it was a moratorium, or seemed to be.

"Yes, but . . ." was how the Uruguayan commissioner to the IWC phrased his support for the proposal. It provoked an outburst of nervous laughter from the extremely tense delegates and was diplomatically recorded as an unequivocal "yes" by the commission's secretary. And yet it was perhaps the single most trenchant comment on the entire proceedings. Yes, conservationists had achieved a *de facto* ban on commercial whaling. But . . .

The announcement came at the end of a long week of hard and delicate bargaining, and gave the campaigners, young and old, justifiable cause for celebration. Joy, if not exactly unconfined, was certainly in evidence as champagne corks popped and scientist embraced activist. The inertia of the IWC's three-quarters majority, so long an obstacle to conservation, had finally been turned in its favour. Whalers, who had always been able to rely on that inertia in the past, would now find that conservationists would be able to use it to block any attempt to resume legitimate whaling.

Therein lies a major problem, at least as I see it. The commissioners of the IWC have always stressed the great store they place on scientific probity. Decisions, great and small, were laid at the door of scientific advice. Political necessities were seldom, if ever, spoken of aloud. But whereas the whalers had always been content to take the benefit of any scientific doubt for themselves, the conservationists had successfully reversed that history and given the whales the benefit of the doubt. The burden of proof was on the whalers.

To do so, they had to compromise their oft-stated principles, a perfect example of the pragmatism—or opportunism—that many conservationists see as one of the main strengths of their movement. In almost every case the majority decisions of the IWC's advisory scientists were ignored by anti-whalers. Where the scientists advised no change, or even increased quotas, the conservationists flexed their muscle and got lower preliminary quotas. Where the scientists offered the whales complete protection, the strategists bargained higher quotas for crucial votes. The moratorium was paid for by selective use of science. And in 1990, when the scientific data may indicate that a harvest may be possible from some stocks, what then will the protectionists do?

Some countries' commissioners saw the problem only too well. Not just Japan, who pointedly drew attention to the more glaring conservationist inconsistencies, but "good-guys" too, like Australia and Antigua. Australia made a policy decision in 1978, that *all* whaling, whether by rapacious profit-motivated factory ship or culturally bound Eskimo, is a bad thing. To stop one and not the other is, as Norway also pointed out, discriminatory.

"There is a greater consideration than the numbers game my colleagues . . . would have us believe is truly the issue," said the Antiguan commissioner, referring to earlier discussions on better ways of killing whales. Humanitarian considerations, not appeals to science, motivated his vote as they motivated Australia's, and that at least is a position that can be consistently defended. "There is no humane method of continuing this needless form of industry," said Antigua.

Events immediately after the moratorium ban was passed are hard to interpret. Japan, Norway, Peru and the USSR all objected, so that the moratorium would not be binding on them. They did not do as well in the next quota negotiations as the countries that had not objected. Brazil, for example, was awarded a quota of 625 Minke whales in 1983, considerably higher than the 454 recommended by most members of the Scientific Committee. Korea, another non-objector, was allowed to extend its quota of Minkes for another two years to take it up to the start of the moratorium. The Scientific Committee recommended a quota of 151 Fin whales for Iceland, but this was increased to 167 animals in the plenary session.

If the good boys did well, the bad boys did not do too badly either. Norway, which had objected both to the moratorium and to the ban on cold-grenade harpoons, got the recommended quota of 635 Minke whales and could have had another 150 but for the reticence of its commissioner. The details of that decision offer yet another glimpse of the inner workings of the IWC.

Norway has consistently sought a higher quota of Minke whales than offered, and the previous year, 1982, the IWC had not been able to agree any figure. This meant that Norwegian whaling was, in effect, unrestricted. In 1983 it was thought that the Scientific Committee's figure of 635 would hold up, but after a series of private commissioners' meetings the Norwegians emerged with a request for 885 whales, and the support of most of the European members of the IWC, including the UK.

Norway's quota was, I am sure, to have been passed by consensus, but conservationists India and Oman called for a vote. The Europeans then felt they had to explain why they would support Norway. In essence this was because Norway was a friendly country and one that was also involved in other fisheries negotiations. Furthermore Peru, also an objector, had been given a quota of 165 Bryde's whales from a much more endangered stock.

Norway spoke before the vote and could probably have got its quota if it had been prepared to give a firm commitment to withdraw its objection to the moratorium. But Commissioner Per Tressalt's statement was pure anodyne, saying only that "Norway was ready to continue to work with other delegations to make it easier for the commission to work with regard to decisions already taken." This could have meant anything, and probably means nothing. At the vote Norway's request for 885 failed to get the necessary majority, despite the support of the UK, and so the original proposal of 635 was adopted.

Conservationists accused Richard Packer, commissioner for the UK, of going against publicly stated policy; in his speech opening the meeting that year Lord Belstead, a minister at the Ministry of Agriculture Fisheries and Food, spoke of "giving the Scientific Committee your full support". But Evelyn Blackwell, alternate commissioner for the UK, told me that "scientific advice has to be tempered with a degree of commonsense and practicality". She was also quite candid about the importance of Norway in other fisheries matters, notably in the North Sea, and further justified the UK decision by pointing out that the Norwegian case "had substance in it purely in relation to the whaling issue".

Many people were nevertheless surprised that the UK actively supported Norway in the vote, instead of abstaining as so many other conservationist countries had. Blackwell explained that her bosses in the ministry regarded the yes vote as "an appropriate way to deal with the particular problems [the Norwegians] had".

The United States successfully extended its block quota for the Alaskan Eskimo hunt of Bowhead whales for another two years.

In 1984 the IWC decamped to Buenos Aires, a slight embarrassment to the UK after the contretemps in the Falklands, but no very important decisions were taken. That autumn, however, the Japanese coastal whalers took several Sperm whales. This was in no sense illegal, since Japan had objected to the IWC ban on Sperm whaling, but it should have opened the Japanese to massive trade sanctions from the United States.

The US has the power to prevent Japan fishing in American waters and to ban fisheries imports from Japan. This represents far more money than the remnants of the Japanese whaling industry. Japan takes 74 per cent of the bottom fish caught by foreign fleets in Alaskan waters, which amounts to about a million tonnes. Even though it represents only 10 per cent of Japan's total fisheries catch it is worth some $750 million, compared to the $40 million of the whale industry. Instead of penalising the Japanese in this much more lucrative market, the United States reached "an understanding" with Japan and declined to impose sanctions, a foretaste of the way in which Japan's adherence to the moratorium was to be handled by the only muscle available (see "Sleazy Deals", p 171).

By the standards of previous years the thirty-seventh annual meeting of the IWC, in Bournemouth, was a disappointment. It should have been exciting, as the final meeting before the moratorium, but it was not. The best news for Save the Whale campaigners, which escaped popular notice, was the USSR's promise that it will stop going after Minke whales in 1987, for what it described as "technical reasons". Cynics said that the Soviet whaling fleet was falling to bits and would have rusted away by 1987.

The Scientific Committee recommended, and the plenary session agreed, to classify Norway's stock of Minke whales as requiring protection status. In effect that means no catch is permitted. During the discussion of this the Norwegian commissioner gave a very good impression of being extremely unhappy with the decision, but I suspected at the time that he was secretly rather pleased. The Norwegians have always set great store by scientific conclusions, but also have to deal with a few bothersome whalers way up north. The Norwegian commissioner now at least had the possibility of arguing that they had to give up whaling because the scientists said so, rather than to comply with the irrational wishes of conservationists. I was wrong, as Norway's continued whaling and the following annual meeting revealed.

Going into the 1985/86 Antarctic season it looked very much as if there might never have been a moratorium. The remaining whalers had objected, and so were not bound by the decision. The Soviets and Japanese sailed down south in force. By the end of the season they had taken 5,569 Minke whales. Japan's private arrangement with the US enabled it to continue taking Sperm whales in the North Pacific. The Norwegians went on with their Minke whale hunt. There was the spectre of "scientific whaling": any country is allowed to award itself a special permit to take any number of whales for scientific purposes, and in the aftermath of the moratorium many supposed ex-whalers expressed a renewed enthusiasm for science. Pirates were most likely active too, somewhere in the world. It was business as usual.

Against this background, the IWC, meeting in Malmö in Sweden, did not vote on a single issue. Instead the commissioners reached a consensus on everything that came up before them. One of the most visible such issues was scientific whaling, which makes sense only because the sale of meat from research whales offsets the costs of the whaling. In an effort to remove this support, conservationists sought to ban all international trade in the products of scientific whales. They failed. Instead, after several alternative wordings had been found wanting, the IWC agreed that meat from scientific whales should be "primarily for local consumption".

The thirty-eighth annual meeting of the IWC saw the traditions of the previous thirty-seven meetings upheld. As with the very first agreements to control whaling, those who stand to profit from continued whaling continue to whale, regardless of any obstacles put in their way. The moratorium has proved no hindrance. One scientist even said that he thought the moratorium would eventually prove to be invisible; when future statisticians pondered the figures for the world's catch of whales they would not be able to detect anything untoward about the years 1985 to 1990, just the same old decline.

"Yes, but . . ." had been the Uruguayan vote for the moratorium. I am reminded of another incident back in 1982 when the moratorium passed. During the celebrations that balmy Friday night an overzealous crewman on one of the Greenpeace boats fired off a red rocket flare. The nearby Shoreham lifeboat immediately put to sea, having (correctly) read the signal as one of distress, not delight.

Science

"Two added to one—if that could but be done,"
It said, 'with one's fingers and thumbs!'
Recollecting with tears how, in earlier years,
It had taken no pains with its sums.

Fit the Fifth: *The Beaver's Lesson*

The International Whaling Commission stands firmly on science. The 1946 Convention says clearly that decisions must be "based on scientific findings". Objective methods are used to assess the whale stocks and tell the whalers how many whales they may safely harvest from those stocks. Time and again in the meetings the commissioners repeat, like a litany, their belief in the power of the scientists and their methods. So in order to understand the IWC we need to understand the science behind whaling.

John Beddington, director of the Marine Resources Assessment Group at Imperial College in London, is one of the world's foremost authorities on the way animal populations wax and wane. He was instrumental in developing the models that the IWC uses to assess the state of whale stocks, and has this to say about the history of whaling: "Possibly the only lasting benefit to mankind from the depletion of the great whales is the extraordinarily detailed data base that their exploitation has produced."[1]

That is a fine sentiment for an academic biologist. The information in the Bureau of International Whaling Statistics in Norway does indeed offer exciting possibilities for investigating many aspects of whale ecology. No doubt someone will get round to that eventually. But for the past few years the data have been put to the much more mundane—but vital—task of estimating the status of the stocks. How many whales remain? And how many may be taken? These were the twin questions of the scientific era of the IWC, and even though the moratorium

[1] Beddington (1980) p 194.

131

has been voted in there is still a need to understand the science of counting whales.

The procedure the IWC adopted, the so-called New Management Procedure, has its roots in a theory developed independently by two fisheries biologists, Wilbur Schaefer and Michael Graham. It takes as its starting point a population of animals that is not being exploited by man. This is the so-called pristine population. Any animals removed from the pristine population will leave space and resources for others. This means that at first the population will reproduce more than it had been before exploitation began. We can take the surplus. As we remove more animals the yield rises to a maximum, the point at which we are removing all the extra animals the population is producing. We are skimming off all the interest while leaving the biological capital at a constant level.

If we harvest only this maximum sustainable yield the population will stay as it is and we will be getting the most from it. If, however, we take more than the maximum sustainable yield we will be eating into our capital. In future years the interest will be lower. The yield still to come will continue to fall as we remove more animals than the population can replace.

The New Management Procedure set a target yield that was equal to 90 per cent of the maximum sustainable yield, erring slightly on the side of safety. In deciding the quotas the IWC needed to know which stocks could be exploited and how many whales could be taken. To do that the scientists needed good estimates of the pristine and current stock levels, and the various parameters describing the rate of reproduction of the whales. And those estimates needed to be accurate, because a 10 per cent change in the estimate of population size could change a stock's status and, theoretically, open or close a particular local whaling industry.

The whalers started to use population estimates to help them manage stocks only in 1959, a decade after the first meeting of the IWC. That was when the committee of three was set up, and the technique they used was based on the notion of catch per unit effort. If there are lots of whales it will take little effort to catch each one, whereas if there are few it will take a lot of effort. For any given level of effort, the number of whales caught will depend on the number of whales there are.

Knowing the catch per unit effort, and how it has changed

over the history of exploitation of a particular stock, the scientists can work backwards to arrive at an estimate of the pristine population size. They can also calculate the recruitment to the population—the increase in numbers thanks to reproduction—with a mathematical model that incorporates details of how many sexually mature animals there are, how many die each year, and how many breed. Armed with data on catch per unit effort and estimates of recruitment to the population they can see how the harvest from the population will change according to the difference between the catch and the recruitment.

All these calculations, however, are founded on one crucial assumption—that the catch per unit effort is directly proportional to the size of the population. The measure of effort is absolutely fundamental to the whole exercise. But the technology of catching whales has not stood still. As whalers moved from sailing ships and hand-held harpoons to steamships with harpoon cannons the effort per whale has obviously dropped. Even so, the scientists say they can make a variety of numerical adjustments to come up with a standardised measure of effort, the catcher-boat day. This produced apparently reasonable data, but in the mid 1970s the scientists, notably Beddington himself, realised that their fundamental assumption was wrong. Catch per catcher-boat day is not a direct proportion of population size. In fact it underestimates the decline in stocks badly.

The reason is that there is more to catching a whale than simply spotting it. The whalers also have to deal with it, and that occupies a period known as the handling time. If all that the catcher boat had to do was see the whale then the number of whales seen each day would indeed be a direct measure of the total number of whales there to be seen. But the catcher boat has to pursue, harpoon, and otherwise cope with its prey. While it is doing that it is not free to go after any other whales, no matter how many the lookouts may glimpse. The handling time is about two hours. So even if there were wall to wall whales beneath the waves, and even if a boat worked flat out round the clock, it could not possibly catch more than twelve whales a day.

When Beddington and his colleagues revised the models to take handling time into account they discovered that the decline in the populations due to whaling had been badly underestimated. There were far fewer whales than they had originally thought. What is even more unfortunate is that the bias is greatest when the actual population level is close to the level for maximum sustainable yield. With the new modelling techniques estimates of stock size had to be reduced, many quotas were scaled down, and a few stocks had their classification changed, from exploitable to needing protection.

The fiddling needed to standardise the whaling effort into units of catcher-boat days had seemed satisfactory, but there was always a worry that the mathematicians were not keeping up with advances in whaling technology. In 1980 the modellers had a chance to try a new method and compare it with the old. The new method used the extensive biological information gained over the years of whaling, and in particular what is known as the full length distribution of the whales.

Whales keep growing throughout their lives, so the length of a whale is a good measure of its age. The distribution of different lengths in a population is thus a mirror of the different ages in the population. The pristine population will have a characteristic length distribution. As the whalers begin to remove animals from the population they distort the length distribution by removing the bigger animals. The degree of the distortion will depend on the size of the catches relative to the size of the population. The scientists looked at the way that the length distribution had changed over recent years and, because they knew the size of the catch, could work out what size population would have suffered those changes in length distribution under that particular catch regime. It sounds a bit complicated, but it works well; applied to slightly older data it produced "predictions" that matched more recent data rather well. The methods worked.

The New Management Procedure drove the scientists to develop more accurate techniques for estimating pristine and current populations. These were used to good effect to regulate some stocks. So it is a peculiar irony that the techniques they developed gave no real insight into the two species then bearing the brunt of the whalers' efforts, the Minke and the Sperm.

The Minke whale in the Antarctic thrived at the expense of its competitors, especially the Blue whale. The Minke was not worth hunting until stocks of the other whales had been severely depleted; it was too small, as long as there were bigger species available. It seems, however, that the removal of so many larger whales from the Antarctic ecosystem created something of a krill glut. What the depleted great whales were no longer eating, the Minkes took full advantage of, with the result that in the mid 1970s there were more Minkes, breeding more rapidly, than there had been before the start of whaling in the Southern Ocean. The number of Minkes appears to have been rising steadily, and the New Management Procedure makes no sense under such circumstances. The current population, when they came to look at it, was probably larger than the pristine population, something not anticipated in the theory of maximum sustainable yield. The New Management Procedure simply could not cope. Instead the scientists tried to get the whalers to exploit Minkes at a level that would keep numbers roughly constant.

The principle underlying this is the so-called replacement yield, which seeks to cream off only the extra whales produced each year, without maximising that yield. Ironically, when a population has been depleted, the replacement yield can be quite high, because the extra resources will sustain extra reproduction. As a procedure for setting quotas, replacement yield ought at least to ensure that no further harm be done to the stocks, but sometimes it offers results that despite their scientific probity are politically unacceptable. That is when the true colours of the IWC members show through their facade of objectivity.

The argument over Minke whales in the run-up to the 1984/5 Antarctic season provides a perfect example. The previous season, 1983/4, the catch limit for Minkes had been 6,655 whales. In 1984 the catch limit was reduced by 37 per cent, to 4,224 whales. Brazil, the USSR and Japan—the only countries taking Minkes in the southern hemisphere—all objected, but as Japan is the marketplace for the other two countries, her objection is the more important. Sidney Holt, anxious that Japan's "superficially . . . plausible" argument be exposed, analysed it in detail, and concluded that the Japanese statement accompanying the objection was "misleading and factually inaccurate".[2]

[2] Holt (1985) p 1.

The primary reason for the 37 per cent drop in catch limits was that the scientists' best estimate of the number of Minkes had been revised downwards, from 405,300 to 258,300, a drop of 36 per cent. Japan could not accept the reduced catch limit because it was not "based on scientific findings". In particular, Japan charged that the decision "was a product of political manoeuvring of some countries", that the Scientific Committee had "simply disregarded" the scientific data in reaching its conclusions, that the Scientific Committee had further been "illogical" in calculating the replacement yield, and that the final figure of 4,224 was based on calculations that were "scientifically unjustified".

The Scientific Committee did not, in fact, "simply disregard" the data. It carefully considered information from marking surveys and analyses of catch per unit effort, and there was a consensus, with no indication of any disagreement from the Japanese scientists on the committee, that such sources were "so variable that no conclusions could be drawn from them". According to its own report, the Scientific Committee "agreed" to base its calculations on sightings cruises undertaken specifically to count the Minke whales. A special study of the catch data from the Brazilian Minkes indicated that they could sustain a larger catch limit than in previous years; the Japanese scientists on the committee strongly opposed the use of those figures. The method that the Japanese objected to was essentially the same as the one they later claimed the scientists should have used. But if Brazil had taken more Minkes, that would have meant fewer Minkes for the Japanese whalers. In any case, the charge that the committee "simply disregarded" the data is, as Holt says, "quite untrue".

The Japanese also said in their objection that it was "illogical . . . that the better stock conditions are, in other words the closer they come to their initial population levels, the lower the catch limits should be". "Illogical it may be," Holt replies, "but this is the inevitable consequence of calculating replacement yields when maximum sustainable yield, required under the IWC formal management procedure, cannot be estimated." Japanese scientists have long recognised this, and have, according to Holt, "always participated actively in discussions of the correct way to perform such calculations, offering useful suggestions on that technical problem". Why, now, were Japanese officials stigmatising a procedure that their scientists had previously found entirely workable?

In fact the catch limits for 1983/4 and 1984/5 were arrived at by very different methods, agreed each time by the Scientific Committee. By chance, the catch limits for the two years were exactly the same percentage—1.64 per cent—of the total estimated number of whales. In 1983 the Japanese commissioner had been happy to be part of the consensus adopting this catch limit. The following year, faced with a quota which was identical in proportion, if not in actual numbers, to the previous year, he voiced very strong opposition within the meeting, and subsequently lodged an objection. What had changed in the meantime?

The reduced quota would perforce cut Japan's whale-meat production by about 3,000 tonnes. Imports from the USSR and Brazil would drop by about 3,500 tonnes. In 1983 and 1984 the Japanese consumed only about 30,000 tonnes, so a reduction of a quarter is not neglible. Set against the previous decade, however, it takes on a different meaning. Between 1974 and 1980 Japan imported more whale meat than it was consuming in 1984, and home production was always greater than imports. The decline forced by the new quota would have been a further small diminution of a practice that had already been hit very hard. The Japanese said that the new quota would "deprive workers of the whaling industry of their very livelihood". In fact the new quota could have been taken by using the same fleet, one factory and four catchers, in five months instead of six, or by using a smaller fleet, three catchers instead of four. "Either way," as Holt points out, "the effect on the required labour force would have been tiny compared with the reductions that have necessarily occurred over the past decade," as the New Management Procedure brought protection to the beleaguered baleen whales.

Holt concluded that the Japanese whaling industry "is an industry which thrived for five decades but which has no significant future, except perhaps in about fifty years time when the whale stocks might have recovered". The Japanese objection, and more especially the justifications for it, which were aimed so squarely at the scientific procedures of the IWC, must be seen as the final throes of a suicidal industry, determined to wring the last pound of flesh from the whales.

Minkes in the southern hemisphere proved impossible for the New Management Procedure to cope with. Sperm whales did too, but presented a different problem. Unlike the other great whales, Sperm whales are polygynous. Big mature bulls hold a harem of several females. Unfortunately, it is just these big bull males that were the whalers' favourite targets. You cannot really blame them, for quotas were set in terms of numbers, not tonnage, but the outcome is that they removed breeding males preferentially from the population. And the consequence of that would be a decline in pregnancy rates, as there would not be enough harem-holding bulls to service all the cows.

Shortly after the New Management Procedure had been instituted there existed a variety of models to describe the breeding behaviour of the Sperm whale. They differed in detail, but all predicted that as the number of breeding males became depleted the pregnancy rate in the females would drop. The modellers were well satisfied when the decline they had been predicting began to show up in the data. Despite this vindication of the models, however, Sperm whale quotas continued to be set, until the ban on Sperm whaling was voted through in 1981. Even now, Sperm whales continue to be killed and the greatest problem is that the scientists simply do not know how the population will respond.

The main bugbear is the long timescale involved, as always in whale biology. Male Sperm whales need to be about twenty-five years old before they are mature enough to hold a harem. So even if catching stops immediately, which it has not, there would be a considerable delay before the population regained its full breeding potential. The stock to the south of Australia, for example, will take decades to recover its former strength, even in the absence of catches. And yet it contains a surplus. That surplus is made up of males that are not big enough to breed.

If the New Management Procedure were rigidly applied it would permit whalers to harvest males. The difficulty is that it is not always easy to tell the sex of a Sperm whale from the bow of a catcher boat. The big whales, the favourite target, are likely to be males, although the IWC schedule forbids whalers to take very big whales, especially during the breeding season. This provision is intended to protect the very largest, harem-holding bulls. The Sperm whale quotas were figured in terms of males, with a percentage set aside for the accidental by-catch of females, but these quotas are very difficult to implement. Its own biology makes the Sperm whale considerably harder to manage than the other great whales.

Perhaps the final irony in the numbers game is that many of the techniques that have been developed operate most reliably on stocks that have been heavily exploited. The mathematical models work best to protect the very stocks whose depletion the analysis is meant to prevent.

Sidney Holt seems now to be quite disaffected by the consequences of the New Management Procedure. He says that the effect of the New Management Procedure "was to close down whaling where there was good information, which in all cases demonstrated depletion, and to legitimise arbitrarily high catches of all other stocks".[3]

Holt is adamant that what he is most certain of is that he does not know enough about whales to manage their harvest. Quite apart from the mountains of unanalysed data from whalers, there are errors inherent in all the other techniques too. Marking experiments, for example, should give an accurate idea of the size of the population. If, for example, you mark 100 whales with a tag, and then find that tag in 1 per cent of the whales you catch, it is reasonable to assume that there are 100 times more whales than you marked, say 10,000. But marking experiments, which have been going on for decades, have been exaggerating the numbers. It seems that the whales can somehow extrude the markers from their blubber, just as they would parasites. Those marks are never recovered, which gives a falsely high picture of the population.

There are other problems too: marks sometimes do not turn up until the whaler looks in the rendering pots. It is then impossible to tell which particular whale they came from, and much valuable information is lost. Marks have their good side though: on a couple of occasions they have trapped whalers who were submitting false records. Marks claimed to be taken from allowed species were in fact originally placed into protected ones. This is small compensation, however, for the one

---

[3] Holt (1985b) p 116.

great fact about the science of whale demographics: it is shot full of holes.

That does not mean it is impossible to manage whale populations effectively, just that one does not need very much science to do so. The recruitment rate is the crux of the matter. If you know how many whales join the population each year, you can get a pretty reasonable quota by simply multiplying the number of whales by the recruitment rate. The problem is that at the moment the scientists' best estimates of the recruitment rate vary from between 1 per cent per year to 4 per cent per year. A fourfold error is not acceptable. And cruises to count the number of whales also produce answers with a wide margin of error. To get around these problems, whale biologists are developing techniques of management based on a system of feedback, a subject that came up at the 1987 meeting of the IWC in Bournemouth.

The feedback approach is actually rather close to real life. You have a rough idea of the yield you can expect, and you modify the quota you set each year according to some set of rules that may or may not depend on how things are going out at sea. If the rules are good ones, you end up taking the maximum sustainable harvest indefinitely. Rather than allowing whalers to go out and try, however, the IWC has agreed to test a series of feedback rules on a model that simulates a real population of whales. The whale biologists have agreed on a mathematical model of a whale population, and can try out various management regimes to see how they affect the whales in the computer, before they risk whales in the ocean. It takes a little while—four hours of computer time for a set of forty trials—but it is much quicker and safer than actual whaling.

By common consent, a good management regime should carry a very low risk of driving the population extinct, it should have catches that are roughly stable from one year to the next, and it should approach the optimal yield level as quickly as possible. With that in mind, Bill de la Mare, Syoiti Tanaka, Japan's foremost fisheries biologist, and a team from Iceland have tried their luck with various management plans. Unfortunately, two of the three approaches do not work.

Tanaka's regime exhibits what de la Mare calls "bizarre properties, such as a noted tendency for populations to go extinct, which is not what you want from a management regime". Tanaka later told me that no scientist could guarantee perfection, but that he was convinced that the feedback approach, of which he was an early fan, can work. De la Mare says Tanaka needs to make his rules more responsive to what is actually going on.

The Icelandic approach does not fare much better, probably because it relies on using information gathered from the harvest to estimate what is happening to the population. "You really need an independent assessment of abundance," advises Sidney Holt.

De la Mare, then, is the only one who can come up with a management regime that works; that is, it does not risk extinction and it does not fluctuate widely from year to year. Unfortunately it also does not offer much potential for profit. That is because it relies on independent sightings cruises and not information from the hunt to inform management how many whales there are, and such cruises cost money. It simply is not worth spending on sightings cruises for the small number of whales that de la Mare's system would safely allow whalers to take now. That does not mean that the feedback approach is useless. "It's probably not economic for Minkes in the Antarctic now," de la Mare told me, "but you could start again with the former abundance. If we went back to the thirties, and learned from our mistakes, we could manage whaling."

So there might be ways to manage whales in the future, without threatening their very existence. There are also techniques being developed to identify whales much more easily and accurately than tags, photographs, voiceprints, or any of the other methods in use today. The most promising of these is called DNA fingerprinting. The name is no mere hyperbole. Every whale carries within its every cell a unique molecular signature, a fingerprint. It takes no more than a few grams of whale tissue and a modicum of biochemical expertise to read the fingerprint, and unlike physical tags the DNA fingerprint can never be lost, altered, or overlooked.

Research scientists are very excited about the prospects of using DNA fingerprints to investigate some of the many remaining mysteries about whales: whether they change groups often, how the mating system works (for unlike human fingerprints, elements of the DNA fingerprint can be traced back through mother and father), a more detailed map of the migration routes. DNA fingerprints could also provide a more accurate estimate of the number of whales than physical marks, but although the scientific prospects are exciting, these studies will not really help whale management. That is because even if scientists use DNA fingerprints rather than tags to do their mark and recapture studies, they will still need to sample a whole lot more whales than they do at present. The accuracy of the estimate depends in large part on the number of animals marked and recaptured. In terms of time and money, putting a sampling arrow into a whale is likely to prove just as expensive as putting in a physical tag.

Scientists are often depicted as an inhuman bunch, cold and dispassionate to a fault. That is as much an exaggeration as all caricatures, and as far as whaling is concerned is untrue in at least one respect. A good deal of research has targeted ways to kill whales more quickly, to make their deaths more humane. The difficulty, of course, is to assess pain in another creature, especially one as foreign to ourselves as a whale. Nevertheless, a speedy end is probably a better end than a lingering one.

The Royal Society for the Prevention of Cruelty to Animals has, as one might expect, focussed its anti-whaling propaganda on the issue of how whales are killed, and a booklet the society published in 1981 makes gruesome reading. It details the injuries suffered by whales in various whaling seasons, when autopsies were carried out to try and establish a final cause of death. Death, especially with the non-explosive cold harpoon, is seldom instantaneous. The first shot may cripple the whale, making it easier for the gunner to get a second "killer" shot in. But as Sir Sydney Frost noted in the Australian enquiry into whaling, it takes a minimum of two minutes to get a second harpoon ready, and often considerably longer, during which time the whale is presumably suffering. "Of the 420 Sperm whales killed by the Australian Whaling Station in 1975, only 195 (46.4%) were killed by one harpoon. An average of 1.7 harpoons was required for each whale."[4] In any case, the killer harpoon is not always lethal. One Blue whale took nine harpoons to kill her and five hours to die.

The time till the whale is dead is another statistic that

[4] Barzdo (1981) p 15.

scientists have attempted to collect, though it is considerably harder to assess accurately than the simple count of the number of harpoons. Most whales seem to die within a few minutes of being struck, the average for Minke whales being just over five minutes in one study, under four in another. These figures are considerably quicker than those recorded in the heyday of Antarctic whaling, probably a reflection of the smaller size of the Minkes, but they conceal another difficult decision.

> Whether these times are acceptable depends on subjective criteria. Clearly a whale that is suffering for 30 mins is suffering considerably but a whale that is suffering for 3 mins may be suffering no less intensively.
>
> In the UK the requirement in domestic animal slaughter-houses is for the animals to be killed with a minimum of suffering. This is achieved by stunning then exsanguination. Cattle killed this way are dead within 17 seconds of becoming unconscious.[5]

The exploding harpoon cannot be considered a humane weapon. As Sir Sydney Frost noted:

> The death of the whale is caused as a result of its organs being shattered by iron fragments from the head of the harpoon. But if the death is not instantaneous, or does not happen quickly, the animal is required to suffer from these truly terrible injuries for at least the three minutes and more usually up to five or seven minutes until a killer harpoon can be fired.

More than one killer harpoon may be required to kill the animal, which then means that its suffering may be increased with each shot.[6] Cold grenades, which do not send shrapnel cutting through the whale, might thus be thought of as an improvement, but they are not. Average times for death by explosive and non-explosive harpoons are similar in Sei whales and Minkes respectively, but the longest recorded death throes are from Minke whales, 30 minutes in Japanese operations and almost an hour in Norway. "This weapon is clearly unacceptable to anyone concerned with animal welfare," says the RSPCA.[7]

[5] Barzdo (1981) p 18.
[6] Barzdo (1981) p 26.
[7] Barzdo (1981) p 27.

The Japanese have done some research into developing a better explosive harpoon for small whales. One experiment assessed the value of the explosive penthrite, but encountered a few problems. Fuses were wont to explode on deck, but sometimes failed to go off in the whale; both events represent an unacceptable hazard to the crewmen. When everything worked, however, killing was instantaneous, or very nearly so, in half the whales, although the average time to death stayed around the three-minute mark. "This was shorter by 22 seconds compared with the . . . previous whaling season during which non-explosive harpoons were used on 3,120 whales."[8]

In search of speedier death, whalers have tried all manner of alternatives. Electric harpoons have been used experimentally and commercially, but the great drawback to these is that blubber is an insulator. The harpoon thus has to penetrate the blubber completely before the current can flow through the whale, and that might not always happen. Nevertheless, electrocution has been used with some success, and is standard practice on Japanese boats when a Minke has not been despatched by the harpoon. Proponents argued that the whale was instantaneously stunned as a result of paralysis of the nerves, "but there is no evidence that paralysis causes unconsciousness. Even in land mammals electrocution is considered cruel if the animal is not first stunned by the current through the brain."[9] To be effective, the power must pass first through the animal's brain. That means putting an electrode on either side of the animal's head, easy in a slaughterhouse but impractical at sea.

The catalogue of methods used to kill whales is nothing if not ingenious. Whalers have tried blowing their prey up with carbon dioxide, fitting a cylinder containing liquid gas to the head of a harpoon. This worked on one or two experimental trials, but would only kill quickly if the cylinder released its gas in the chest cavity. Anywhere else, and death would come as slowly as from a conventional harpoon. Again, the animal would not necessarily be unconscious as it died.

Basques tipped their harpoons with bottles of sulphuric acid. Prussic acid and hydrogen cyanide found favour towards the

[8] Anon. (1982) p 6.
[9] Barzdo (1981) p 20.

end of the nineteenth century, but took half an hour to kill the whale and did not fill the crew with confidence. Curare, aconite and other poisons have all been used at one time or another, and there is the old Norwegian method of infecting the whale with pathogenic bacteria and letting disease do the rest. Ingenious though these are, they can hardly be described as more humane than Svend Foyn's harpoon.

Aboriginal methods are also clearly inhumane, death taking an agonisingly long time. Some whalers, like the Alaskan Eskimos, have tried to improve their methods. Others, like the Faroese who herd pilot whales into small bays before gaffing and slaughtering them, cling to their old ways more as a mark of independence than necessity. But then, aboriginals are exempt from most of the considerations of the IWC, and it is a brave person, like Sir Sydney Frost, who would impose "civilised" moral standards on aboriginals.

Cruelty, despite all the research, has never been a prime concern of whalers. Speedier despatch generally means less risk and speedier processing, and therefore more whales. Perhaps if the humane aspects of whaling were more widely appreciated the industry would have stopped sooner, but such considerations carried very little weight. The RSPCA's conclusion, for what it is worth, is that "death or unconsciousness should be instantaneous. In the developed world it is entirely reasonable that if whales cannot be killed humanely they should not be killed at all."[10]

Science supposedly held all the trump cards in the run-up to the moratorium vote at the IWC's meetings. The Scientific Committee and various sub-committees deliberate for weeks before the commissioners begin their haggling, attempting to come up with recommendations to pass up the chain. The difficulty is that the science of whaling is imprecise, so there is always room for manoeuvre. Seldom do recommendations emerge from the Scientific Committee with the full backing of all the members, unless there is absolutely no interest in the point. And the machinery of the IWC is such that the report of the Scientific Committee, with all its dissenting opinions, gets to the full meeting of commissioners only after being filtered through a

[10] Barzdo (1981) p 31.

meeting of the so-called Technical Committee. There is ample scope for disagreement, and with a few exceptions it seemed that many of the scientists attending the meetings were the puppets of their political masters.

Each appeared to go into the meetings determined to insert into the report a form of words that his commissioner could use later in debate to bolster political or economic desires. Where there are legitimate differences of opinion, as there are bound to be in a subject as fraught with unknowns as whale management, the scientific reports often contain contradictory advice. In these cases political decisions must be made, and yet neither the scientists nor the commissioners would admit that this was so. They present to the public a concerned and rational face, yet behind that face they must be aware that their decisions are anything but scientific. Only occasionally does the mask slip, as it did in 1986.

A new committee of four surfaced at the 1986 annual meeting: Doug Chapman, Sidney Holt, Roger Payne and Bill de la Mare, with more than sixty years' combined experience of the IWC, were intent on drawing attention to the North Atlantic stock of Minke whales. The four banded together in their capacity as scientists, rather than as national delegates to the IWC, to issue a strong statement of their "deep concern about the impending fate of the Minke whale".[11] The data for catches in the North Atlantic, almost exclusively by Norway, are very poor and demonstrably unreliable, and that was one thrust of the statement. Despite these shortcomings, however, the four scientists were confident that the numbers of Minke whales "have been declining for many years ... at an accelerating rate". There may be room for some scientific disagreement, they admit, but as with previous unresolved issues they wanted the commission now to give the benefit of the doubt to the whales. "Scientific uncertainties have been used as a convenient means of maintaining advice for catch quotas too high and allowing hunting to continue for too long," they asserted. "This is a serious abuse of legitimate scientific disagreement."

Per Tressalt, the Norwegian commissioner, took all this in his stride. In previous years he had always set great store by

[11] Chapman et al. (1986).

146

scientific conclusions. For example, in 1982 the Scientific Committee recommended a catch limit of 1,690 Minkes. Conservationists attempted to push through a lower figure, but as a result of an organisational snafu no quota for the Norwegian Minkes was set. Tressalt promptly gave the IWC "an assurance that in the absence of any quota the Norwegian catches will not exceed 1,690".

Now, however, faced with the IWC's decision, based on the recommendation of the Scientific Committee, to give the North Atlantic stock of Minkes full protection, Tressalt's government had objected and Norway's whalers had set out to continue their pursuit of the Minke. At the IWC, Tressalt told the Technical Committee in no uncertain terms that "the objectives of any management programme, whatever the resource, will always have a political nature. There is nothing in science which can guide us to what our management goals should be."[12] Sir Peter Scott later quietly volunteered me the opinion that "Tressalt is a nice man, and an excellent diplomat, but he is a trifle short on integrity."

Another prime example of the tempering of rational principles with political pragmatism is the way that the issues of the Blue whale unit and the moratorium were approached. When the scientists wanted to abolish the Blue whale unit, because it did not deal with each stock according to its needs, the IWC baulked. There was no reason to abandon the Blue whale unit, the commissioners said. When the scientists wanted to protect all stocks together under a blanket moratorium, the IWC again baulked. There was no reason to return to lumping all stocks together. The scientists had changed their minds, as good scientists always say they will if the facts demand it. But the negotiators could not cope with that.

I could easily document scores of occasions on which science was followed or ignored, by commissioners who inevitably claimed that their vote was motivated by rational concerns, but it would serve no purpose. The simple fact is that although the IWC is supposed to base its management on science, it failed for a long time to adopt the suggestions of the majority of its scientists.

---

[12] Technical Committee meetings are normally closed to journalists. A fault in the loudspeaker relay allowed me to hear Tressalt's remarks.

In the early days of the scientific IWC the decisions generally favoured the most optimistic scientific conclusions. Quotas were usually at the high end of the range. More recently there has been a shift to a more cautious attitude, although this is still not accepted by die-hard whalers. Not that they disagree with the principle of scientific management of whale stocks; just with the particular recommendations that emanate from the most respected scientists.

When all the room for error is taken into account I think one has to agree with Holt: all we really know about whale reproductive biology is how little we really know about whale reproductive biology. It has taken a long time to take seriously the scientific argument that ignorance is as good a reason for a moratorium as certainty. And even where the scientists were quite certain, and quite unanimous, the IWC did not always listen.

Bowheads

Down he sank in a chair—ran his hands through his hair—
 And chanted in mimsiest tones
Words whose utter inanity proved his insanity,
 While he rattled a couple of bones.

Fit the Seventh: The Banker's Fate

One whale, more than any other, symbolises the plight of all. Its history is the history of commercial whaling. Its protection could give meaning to the continued existence of thousands of people, and yet attempts to protect it have resulted in unnecessary deaths among other whales. It is the Bowhead.

Now reduced to less than a tenth of its former population, the Bowhead is the only whale threatened with actual, rather than commercial, extinction. Scientists of the International Whaling Commission are agreed that further hunting poses a severe threat to the Bowhead as a living species. Indeed, they cannot be certain that the Bowhead will not continue in a terminal decline even in the absence of any further hunting. The Bowhead, however, is taken by Alaskan Eskimos, and has been for the past millenium. It is the fulcrum of their culture. Thus, despite the unanimity of the scientific advice, the United States government has been caught in a cruel and compromising bind, a bind made tighter by the fact that it is the only government with the muscle to impose unwelcome decisions of the International Whaling Commission on reluctant members. The US wants all other nations to cease commercial whaling, but it needs to protect the rights of one group of its citizens to carry out a traditional whale hunt. The diplomatic horse-trading (sell-out might be a more appropriate term) necessitated by this dilemma has not always been publicised by otherwise zealous conservationists, and yet the story behind the Bowhead story provides a revealing insight into all aspects of whaling.

The Bowhead is one of the Right whales: it is large, swims slowly, and conveniently floats when dead. (Not every

Bowhead floats when dead, a fact that has interesting repercussions for the current Eskimo hunt, but enough of them do stay on the surface to make them easy to hunt.) That is why the Bowhead, and the closely related northern Right whale, were the whales to watch for. They are big, too, about 15 metres long when sexually mature, and weigh about 90 tonnes. Exceptional specimens yield nearly 33,000 litres of oil and 1,500 kilograms of baleen. There is no fin on the back, probably an adaptation to a life spent partially beneath Arctic ice. The tail flukes are large, and deeply notched, often displayed to perfection as the whale sounds on a deep dive.

The Bowhead gets its name, not very originally, from the shape of its head. The jaw is arched upwards like a bow, and in adults the head can be as much as two-fifths of the body length. Inside the jaw, accommodated within the arch, are the baleen plates, the longest of any whale species. In the early days of the industry plates more than 5 metres long were not uncommon, but in modern times the average is between 4 and 4.5 metres. The baleen is dark grey or black, but in the sunshine it often appears to glitter with a greenish metallic sheen.

The huge head, and the baleen plates within, enable the Bowhead to pursue its particular way of life. It feeds by swimming slowly at the surface with its mouth open, trapping all the organisms of the plankton in the fine fringes of the baleen. From time to time it will dislodge the accumulated food with its enormous tongue, or else by shaking its head and producing a noise, the baleen rattle, that can be heard over long distances. It swallows the mass of food with what is apparently quite an effort, and returns to skimming its sustenance from the water.

Food is particularly rich in the more northerly waters of the Arctic, where deep ocean currents and rivers draining off the land bring nutrients to the surface and support the flourishing plankton. The Bowhead moves back and forth each year with the ice, staying close to the edges and occasionally breaking through ice up to a metre thick in order to take a breath. The whales calve in the spring as they are migrating north to the feeding grounds, and the young may have been weaned by the time they come south again in the autumn.

Along with the northern Right whale, the Bowhead was the mainstay of the first commercial whale fisheries. There were at least five separate breeding stocks of Bowheads, each in turn ravaged by the whalers.

First, the Basque fishermen took Bowheads in the Bay of Biscay. Then, in the early years of the seventeenth century, Dutch whalers sailed up to the waters off Spitsbergen to exploit the Bowheads there. The Dutch shared the spoils with the much less successful British, and within 200 years the original population of about 25,000 had declined to the point of being commercially unprofitable. It is probably extinct in these waters, only twenty-three having been seen since 1958. The Davis Straits, between Greenland and Canada, and the Hudson Bay also supported intensive whaling for Bowheads. Some 6,000 were taken before 1915, and the population now is believed to be in the low hundreds. There may be fewer than a hundred left.

The remaining Bowhead populations are on the other side of the North American continent, one in the Sea of Okhotsk, north of Japan between Kamchatka and Sakhalin Island, the other migrating between the Beaufort and Chukchi Seas through the Bering Straits to the Bering Sea. It is this latter, the Western Arctic stock as it is called, that is of primary concern; man's interaction with these animals has passed through four distinct phases.

Eskimos and their ancestors have been hunting these whales since at least 500 years before Christ. Brave hunters, in skin-covered boats, took perhaps sixty whales a year, using toggle-headed harpoons and inflated sealskin floats on lines attached to the harpoon. The swivelling toggle head anchors the line firmly in the blubber beneath the whale's skin, while the sealskin floats act as drags, tiring the whale.

The Bowhead was the hub of all life for those who hunted it. "Whaling is our Christmas, Fourth of July, and Thanksgiving" was how one crewman described its significance—religious, cultural and nutritional. The hunt is enmeshed in ritual. Magic songs make the whale swim more slowly, or tire more quickly, or turn back towards the shore. The captain's wife, left behind at the village, must not comb her hair until the men return. Around the house she must move slowly and quietly, for the whale will mimic her when the hunters strike it. There is a mystical link between the harpooner and his quarry, as one old Eskimo whaler explained:

> They say, some men, when they harpoon the whale, the whale goes kind of weak, slow. And some men, they harpoon the whale, they say it's real swift, fast. They say the harpooners are two kinds of people. One crew . . . every time when they strike a whale . . . it turns back. This happened several times. As a matter of fact, I crewed with him about three springs and every time when we struck a whale, it automatically turned back.[1]

It doesn't matter whether this is true or not, the real truth is that whales are central to Eskimo culture. The whaling captain is an important man, tied to his village by rights and obligations that include distributing the catch to all members of the village, and it is an extremely valuable catch. An average Bowhead is an enormously abundant resource to the Eskimos, providing them with about 50 tonnes of raw materials and food. In especially fruitful areas the Eskimos could gather half of their winter food supply in just eight weeks of work; the Bowhead became a central part of Eskimo culture "both calorifically and spiritually"[2] and the surplus it provided enabled rich artistic, religious and mythological traditions to develop. This aboriginal co-existence constituted the first phase of Bowhead exploitation.

The second phase of Bowhead whaling began in 1848 when a Yankee whaler called Captain Thomas Roys, out of Sag Harbor on Long Island, discovered the happy hunting grounds of the Arctic Ocean. He filled his ship with 1,800 barrels in less than a month, taking 100 barrels (12,000 litres) of oil and a tonne of baleen, or whalebone, from each Bowhead. Roys bruited his discovery about on Hawaii and set off a veritable stampede, a new oil rush; in 1850 more than 200 whaling ships cruised the icy waters of the Arctic Ocean, hunting the Bowheads. By the late 1860s catches were already declining, and the whalers turned to walruses while waiting for the whales to come south again from their feeding grounds high in the Arctic.

---

[1] *Umealit: The Whale Hunters* (1980).
[2] Bockstoce (1981) p 166.

The decade around 1880 marks a watershed. Between 1875 and 1885 the price of whale and walrus oil dropped precipitously, largely as a result of increasing supplies of petroleum and fish oils. At the same time the price of baleen, demanded for fashionable skirt hoops and corsets, not to mention buggy whips, household brushes and other domestic goods, rose as steeply. The captains of the whaling industry, already faced with falling catches, responded with steam-driven ships, better able to manoeuvre in the ice and thus pursue the whales. They were rewarded, for a time, with a bigger harvest, but were still missing out on the greater portion of the Bowhead population, which travelled north along the Alaskan coast in spring. The whalers were seldom able to get through the retreating pack ice in time to be in a position to capture these whales, so they gave up their sea-going vessels and adopted the Eskimo methods; the whalers got the valuable baleen, the Eskimos most of the meat and blubber.

Between 1885 and 1900 American whalers established nearly twenty shore-based whaling stations to intercept the whales on their springtime migration. The stations hired Eskimo crews, introducing them to the advantages of Western killing technology. Where the Eskimos had used bone-tipped lances and hand-thrown harpoons, the whalers gave them darting guns, thrown like a harpoon but with an explosive bomb at the tip, and shoulder guns, which fired a similar bomb through the air. It was not all new technology, though; the Yankees saw the superiority of the Eskimo toggle heads and float drags, and were quick to adopt them. The whalers also sucked the Eskimos into a cash economy for the first time. The shore-based whalers processed about 1,000 whales between 1880 and 1910.

A final burst of Bowhead whaling took place between 1890 and 1910 in the Beaufort Sea near the delta of the Mackenzie River. This followed a report by an American whaler, who had spent a summer in the area and saw large numbers of whales feeding there. A big fleet of steam whalers set out to confirm his story, and for a couple of years took a great many whales, but they were so far from their home port of San Francisco that they had to spend the winter in the Arctic. This, together with changes in the supply of whales in the Arctic and the demand for whale products in the markets back home, brought an end to the second phase, of all-out exploitation.

According to John Bockstoce, curator of the Whaling Museum of the Old Dartmouth Historical Society in New Bedford, Massachusetts, it was the very scarcity of the Bowhead at the end of this phase that saved the whales from ultimate extinction. In the sixty years of the fishery some 30,000 Bowheads were taken from Arctic waters. At the end of that period the price of whalebone had risen to $5 per pound, which valued a full-grown Bowhead at some $10,000. The market was ripe for a substitute, and with the introduction of spring steel the value of a pound of whalebone plummeted to less than 50 cents within three years. In 1907 whalebone sold for $5 per pound; in 1912 the price was 7½ cents. There was no longer any profit to be had going north in search of Bowheads, and almost all the shore stations had closed by 1910.

For the whalers this was bad news, although it was mitigated by the increased profits to be had down south in the Antarctic. For the Eskimos, whose numbers had grown, and who had become dependent on the whalers' cash economy, it was a disaster. They were forced back to subsistence hunting, the third phase of the Bowhead saga.

Between 1910 and 1965 the Eskimos went in pursuit of the whale more or less as they had done before the Yankee whalers arrived. They retained the more efficient shoulder guns and darting guns, and about thirty-five whaling crews set out each year. Mostly they were unsuccessful, not surprising when you consider that their resource had been devastated. Early on they landed about ten whales a year, compared to about sixty a year before 1848. This limited take was presumably within the natural productivity of the whales because, without any increase in effort, the catch climbed slowly to reach about fifteen per year in the early 1960s.

During this second subsistence phase, the captain of a whaling crew regained his pre-eminent position in Eskimo society. With equipment scarce and expensive, and with no great cash value attached to the whale, it was very hard for a young man to become a whaling captain. He had three options: he might inherit his whaling equipment; or he might marry into a family that had no son to inherit; or he might try and amass the great wealth needed to purchase a new set of whaling gear. The 1960s saw the third route become ever easier. The re-emergence of the cash economy, based largely on the boom in exploitation of mineral resources, allowed the most unskilled labourer quickly to earn the money needed to equip a whaling crew. The new Alaskan bonanza ushered in the fourth phase of Bowhead whaling.

By 1976 there were two or three times as many crews active as there had been in 1960. Worse, those crews were demonstrably less competent. Catching a Bowhead is never easy. Sometimes whales are killed but cannot be landed for butchering on the ice. Often they are struck and lost, either under the ice or out at sea. Many of those struck and lost die. The experienced whaling captains, who had served a long apprenticeship, occasionally failed to land a whale, but the new breed of cash-rich captains lost whales much more often than their predecessors had. In 1974 twenty whales were successfully killed and butchered. Three were killed but lost. At least twenty-eight were struck and lost. In 1976, forty-eight were butchered while forty-three were struck and lost. By the time of the International Whaling Commission's annual meeting of 1977, with the autumn season still to come, twenty-nine Bowheads had been killed and butchered. Three were killed but could not be landed. A further seventy-nine had been struck and lost.

The International Whaling Commission had not been unaware of the plight of either the Eskimos or their whales. The 1931 convention, while offering the Bowhead complete protection against whalers, who were really no longer interested in it, had also granted a special exemption to the subsistence hunt of the Eskimos. This exemption carried forward into the IWC convention of 1946. The increase in the hunt in the 1970s did not escape the Scientific Committee of the IWC, which in 1972 started calling on the US government to provide better information on the Bowhead and the hunt, and to take steps to reduce the number of whales struck but not landed. The US did almost nothing. Nor did it transmit the IWC's concerns to the Eskimos. In 1976 the Scientific Committee called for a proper study of Bowhead numbers now and in the past, and urged the United States to "limit the expansion of the fishery and reduce loss of struck whales (without increasing total take)"[3]. Again,

---

[3] Dudley and Gordon Clark (1983) p 13.

nothing was done. At the 1977 meeting the IWC was faced with evidence of a hugely increased hunt, 111 animals. As Ray Gambell, secretary of the IWC, explained, "The numbers being removed were absolutely catastrophic. You just cannot take a hundred whales out of a population which you think is only a thousand strong and hope that it will survive."[4] The IWC voted to remove the Eskimos' exception. No Bowheads were to be taken in 1978.

All Hell broke loose. Thanks to the United States' inactivity the ban, even though it had been on the cards for the previous five years, came as a shock to the Eskimos. They charged that the government had "abrogated their inalienable native hunting rights". The government is said to have offered free beef as a substitute, which would feed the Eskimos and prevent any compensatory depredations on the caribou herds. The Eskimos retorted that food was only half the problem; so important was the whale to Eskimo life that without it they would suffer cultural starvation.

The US was trapped between its previously strong stance in favour of whale conservation and President Carter's support for human rights, especially those of minority groups. As a member of the IWC, the choices facing the US were stark. The government could abide by the decision and put an end to Eskimo whaling; or it could lodge an objection and forgo any standing it might have enjoyed with conservationists. Carter decided in favour of the whales. There would be no objection, but as the ninety-day deadline approached the US indulged in the first of many subsequent pieces of consummate brinksmanship.

First the newly formed Alaskan Eskimo Whaling Commission persuaded Judge John J. Sirica (he of Watergate fame) to order the State Department to file an objection. On the eighty-ninth day the State Department had this ruling reversed by the Circuit Court of Appeals in Washington. The AEWC attempted in its turn to appeal against the appeal, but with literally hours to go the Chief Justice of the Supreme Court, Warren Burger, refused to consider their document. The government then put forward its own solution: a quota of fifteen whales killed or thirty struck, in concert with a crash

[4] *Umealit: The Whale Hunters* (1980).

programme of research into Bowhead biology. The IWC, at a special meeting in December 1977, rejected the quota offered, agreeing instead to limit the Eskimos to twelve caught or eighteen struck.

In 1978 the Eskimos stayed within the quota and the US presented its first quick and dirty estimate of whale numbers —2,264. The IWC rewarded the US by increasing the quota to eighteen landed or twenty-seven struck. This was still not enough for the Eskimos, who walked out of the IWC meeting. They announced their firm intention to ignore the IWC quota; bad weather thwarted those intentions and, through no fault of their own, the Eskimos stayed well within the quota, striking twenty-two whales and landing seven.

Early in 1979 an extraordinary meeting of international specialists in Arctic biology, anthropology and nutrition concluded that "although there was no credible scientific basis for allowing any Bowhead catch, and although alternative sources of food are available, whaling is so culturally important to the Eskimos that it should be allowed to continue at some level." Since then, the Bowhead quota has returned again and again to haunt the US delegation and thwart efforts to save other whales.

In 1979 a special sub-committee of the Scientific Committee reported that "the safest course for the survival of the population is for the take to be zero".[5] Despite this, the Technical Committee offered eighteen landed or twenty-seven struck once again. Derek Ovington, the Australian commissioner, reminded the Americans of President Carter's pledges on whale conservation, saying that "we cannot ignore the real danger that this species is on the brink of extinction. The IWC will have little credibility left if it for the fourth time ignores the Scientific Committee."

Dick Frank, the US commissioner, entered some very special pleading. "A vote for a zero quota in this case is a vote against aboriginal people," he said, and the proposal for a zero quota failed. Sorrowfully Frank admitted that he could not tell the Eskimos that anything less than eighteen was fair, and made veiled threats that he might not be able to enforce a lower quota. But he did offer to give up one struck Bowhead.

[5] Anon. (1979) p 3.

Australia at this point struck a low blow. "The distinguished delegate from the US has suggested that Spain control the operations of the *Sierra*," he said, alluding to the infamous pirate ship owned by South Africans, flying under a succession of flags of convenience, notably those of Cyprus and Somalia, and catching whales for the Japanese market, as it happens in Spanish waters. "This is in contrast to the US, who cannot even control its own citizens."

The US finally got the Bowhead quota it needed. Overall, that year, the quotas totalled 15,835 whales, 467 fewer than in 1978. Earlier in the meeting there had been a strong proposal to ban all commercial whaling. It was watered down to the partial ban on some factory fleets, which affected only the activities of Soviet fleets pursuing Sperm whales. The full commercial ban had originally been tabled by the United States.

Reaction in some quarters was very strong. Dr Lee Talbot, at the time Conservation Director of the World Wildlife Fund, was particularly vehement: "The quotas would have been a lot lower if it hadn't been for haggling over Bowheads," he told me. "We have a new Blue whale unit. It is called a Bowhead, and it is equivalent to about 146 Sperm whales."

Certainly 1979 showed clearly the profound difference in attitude between the ethically committed Australians and the politically expedient Americans. The following year, 1980, was even more eventful for connoisseurs of horse-trading. Further studies had demonstrated that the Eskimos "needed" thirty-two or thirty-three whales to satisfy their nutritional requirements. The 1979 quota did not meet these needs. The scientists had not come up with any startling new research findings; all they could do was reiterate their concern and warnings. As Dick Frank expressed it for the benefit of journalists: "Scientific analysis indicates that the population is likely to decline even in the absence of a harvest, but will decline more rapidly if the current harvest levels were continued for an extended period of time."[6]

The dilemma had become apparent to even the meanest of intelligences. Allow the Eskimos to kill Bowheads and they would put the whales, and their own culture, in jeopardy.

[6] Mark (1980) p 1.

Prevent the Bowhead hunt and you interfere with the way of life of an important minority group, a policy no longer favoured by the US government.

Lyall Watson, alternate commissioner for the Seychelles, recognising the acute embarrassment caused to the conservationist US by its need to obtain a Bowhead quota, tried to shift the matter up the agenda, so that it could be settled before the debates on a moratorium, on humane killing, and on the quotas themselves. It was a nice try, but alas it did not come off. Technical Committee's recommendation—zero Bowheads— went down smartly, so Watson made a second effort. He proposed a quota of eight whales, one for each of the villages on the Beaufort Sea coast, a kind of ritual sacrifice. That too sank, after the US had put in an impassioned defence of an unaltered quota. The UK proposed an adjournment, and the Bowheads spent the next four days in smoke-filled rooms and corridor conclaves.

On the last night of the meeting they emerged; the US was offered a total of forty-five whales landed (or sixty-five struck) over the coming three years, with a proviso that no more than seventeen be landed in any one year. The vote sailed through, as indeed it ought to have, given the thoroughness with which its passage was presumably charted.

In the meantime, other decisions had been taken. The commercial moratorium, again on the agenda, had again failed. A more limited ban, to protect only Sperm whales, but this time completely, failed by just one vote, probably Canada's. Minke whale quotas were reduced, but not by as much as the majority of the Scientific Committee had wanted. Iceland's take of Fin whales was not altered, despite a Scientific Committee recommendation of total protection for this stock. Spain, which had only recently joined the IWC, obtained an increased allowance of Fin whales in the teeth of the scientific advice. Japan held out for a quota of 890 Sperm Whales (the scientists recommended considerably fewer), and got it.

Talbot's Bowhead whale unit was still going strong, and it is no exaggeration to say that aboriginal needs allowed commercial whaling to continue. But the block quota signalled the beginning of a new stage, and pragmatically was probably a good tactical decision. The IWC has always been more concerned with preserving whales for commercial purposes than with saving them from oblivion. The Bowhead is facing extinction, but it has not been of commercial importance since about 1910. Dick Frank's limited success, coupled with the IWC's built-in voting inertia, meant that the Bowhead problem could be shelved for a couple of years at least. That freed the US to use its considerable muscle in a push for a complete commercial moratorium.

So it turned out. In 1981 the great Sperm whale ban was passed, although it has yet to stop Japanese whalers from killing Sperm whales. And the complete moratorium forced its way through the required three-quarters majority the following year, 1982. That too has yet to bite. In 1983 the Bowhead quota was once again up for negotiation.

Long before the July meeting of the IWC the new US commissioner, John Byrne, entered into a formal agreement with the Alaskan Eskimo Whaling Commission, to get them a quota of thirty-five whales. This angered some conservationists, who claimed that Byrne had failed to consult other interested parties. The Scientific Committee for the first time showed some evidence of its awareness of a real world outside the committee room; its report said that it would still prefer no Bowheads to be taken, but if a quota had to be set it should be no greater than twenty-two animals struck, with an emphasis on animals less than 13 metres long. (These would be sexually immature, so harvesting them would not affect the growth of the population too much as their parents would be left to continue multiplying.)

The scientists had been continuing to refine their estimates of the Bowhead population, and now had a better idea of its rate of increase. The figure of twenty-two was based on that. Byrne admitted that his figure of thirty-five was based on "fuzzy science" and that the method of arriving at it—never fully explained—was "quick and dirty". In the end he came away with a two-year block quota of forty-three whales, with no more than twenty-seven in any one year. For the first time the quota made no distinction between whales struck and whales landed, which should have been an incentive to the Eskimos to improve their efficiency.

With the Bowheads once more out of the way the 1984 meeting, in Buenos Aires, saw little mention of these beasts, but in 1985, in Bournemouth, they were back on the agenda. In the meantime there had been two notable developments: first, the United States had entered into a series of private deals with the Japanese, permitting them to continue whaling for at least two years beyond the start of the moratorium; second, the Bowhead population, 2,264 in 1978, had increased to 4,417.

Well, not quite. In fact the number of Bowheads had probably not changed much at all, but the scientific estimates of the number had. The researchers had been working hard to improve the accuracy of their census methods. In the early days they simply sat on a perch by the ice edge and counted the whales swimming past. If they were able they used two observers at different perches, which gave a better estimate. But it was clear that some whales were evading the count. They might be swimming too far away to be seen from the perch, a fact confirmed by several aerial surveys. They might be swimming past in bad weather, when visibility was low. This too was confirmed using an array of underwater microphones to follow the sounds of the migrating whales. The acoustic survey also revealed that whales sometimes swam under the ice, thus avoiding the researchers' gaze. With these new methods the scientists discovered that something like half of the whales were heard but not seen.

All of this additional research threw up several correction factors that were plugged into the equations, allowing the scientists to state that they were 95 per cent certain that in 1978 the number of Bowheads had been somewhere between 2,909 and 3,971. By 1984 the figure lay between 2,613 and 6,221. (Note that there is still an awful lot of slop in these figures. The average may have gone up from 3,440 to 4,417, but such an increase is biologically highly unlikely and there is a reasonable chance that the total population in fact went down.) The Scientific Committee remained unable to predict with any confidence the likely trend in Bowhead numbers. They simply could not say whether Bowheads were increasing, decreasing, or staying the same. As a result they were unable to recommend a quota, once more urging caution instead.

The United States still wanted thirty-five whales a year, but the proposal sent forward from the Technical Committee offered a continuation of the previous quota, forty-three strikes over two years with a maximum of twenty-seven in any one year. It went slightly further, however, noting that if the figures to justify an increase could be produced the IWC would re-examine the Bowhead quota after just one year. The US was unhappy about this. "It is important to recognise the responsibility and performance of the Alaskan Eskimos," said Commissioner Byrne. "We are sincerely confident that the request for 35 strikes is not an overly large request." The Technical Committee's two-year block quota failed, and once again there was an adjournment to see if a deal could be struck and landed.

On resumption of the full session of the IWC the chairman proposed a way round the impasse. "For each of the years 1985, 1986 and 1987, 26 whales may be struck; however, strikes not used in any one year may be carried forward." This quota was subject to a limit of thirty-two in any one year, and also allowed for a reassessment of the ceiling of twenty-six each year. It neatly increased the 1985 quota from the eighteen left over from the previous two-year allowance, and it came closer to the Eskimos' stated needs. It sailed through on the nod, with the exception of Mexico, who boldly reserved her position.

Technical Committee had recommended twenty-seven whales a year (ignoring the fact that if this number were taken there would be just 16 available the following year). The Eskimos wanted eight more. Just before the vote on the Technical Committee's proposal the chairman of the Alaskan Eskimo Whaling Authority, Lennie Lane Jr, was allowed to address the IWC. He said:

> We need 35 strikes to keep our ability to be responsible resource managers. We have to show our people that we are keeping our faith with them. We cannot do that if the IWC denies us eight whales for no good reason. Our people know that eight won't hurt from 300 born. Don't, for the sake of eight whales, cause us to lose all that has been gained in the last few hard years.

It was a moving speech. It prompted many countries to concede that they would have no problem with giving the Eskimos eight additional Bowheads. The US delegation, worried about the consequences, sought to stifle this compassion. "If it will help the confusion," said Byrne, "this proposal is not one the United States will support and so we will vote no." Other countries took the hint, and in the event the Eskimos got not eight extra but five. Two months later Lennie Lane Jr had committed suicide.

The Bowhead whale may yet survive, may yet provide the Eskimos with the cultural underpinning they need for another thousand years. Its history, however, does not give grounds for optimism. When the Eskimos developed their techniques—not for nothing do they call themselves The People of the Whale— they lived in harmony with the Bowhead. It provided for their material and spiritual needs, and could have continued to do so forever. With the advent of commercial whaling the Bowheads, and the Eskimos, were exploited almost to the point of collapse. Now, with whaling in decline and Eskimos in a partial ascendancy, the Bowhead is again threatened.

There could, in fact, be an even bigger threat ahead. So far, government and Eskimos have been relatively happy with the results of their research programme, which has continually upped the average number of Bowheads. But they have recently had some scary news. The average age of the Bowheads may be much greater than scientists had assumed. Donald Schell quietly dropped this bombshell at the Fourth Annual Conference on the Biology of the Bowhead Whale, held in Anchorage in February 1987. Schell is a chemist at the University of Alaska in Fairbanks, and he has made a careful study of the carbon isotopes in Bowhead whale baleen.

Isotopes are different varieties of the same chemical element. The balance between two isotopes of carbon depends on temperature, and the baleen preserves an almost perfect record of the temperature of the water in which the Bowhead has been feeding. Schell discovered definite cycles in the ratio of two isotopes, C^{12} and C^{13}, as he sampled along the length of the baleen plate, and thought these might reflect the annual migration between the icy Beaufort Sea and the warmer Bering Sea. To confirm that they were annual cycles, Schell looked at another isotope, C^{14}. In 1963, the year before the partial test-ban treaty, the US and the USSR exploded 550 megatonnes of nuclear weapons in the air. That more than doubled the amount of C^{14}, which has been decaying since then at a known rate. So Schell could measure the cycles against the amount of C^{14} left. They were annual. And there were far more cycles than anyone expected. That meant that the whales were far older than previously assumed, which has far-reaching implications for the Eskimo hunt.

The animals the Eskimos catch at the moment fall into two distinct size classes, one about 8.5 metres long, the other about 10.5 metres long. Biologists have generally assumed that these whales are one and two years old, and the IWC has based the quotas on this assumption. Schell's isotope calendar clocks the one-year-olds as actually six, and the two-year-olds as ten. Worse still, Schell does not think the animals become sexually mature until they are at least fourteen, and the IWC model needs them to start breeding at between four and six to keep the stocks up. Even the figure of fourteen could be an underestimate because the baleen fringes fray and some of the annual cycles may have been worn away.

Exactly what these revelations will signal for the Bowhead quota is hard to foresee. There has already been angry reaction to Schell's results, but he points out that he was only bringing hi-tech to bear on some observations made by William Scoresby, the English whaling captain of the nineteenth century. In 1820, Scoresby pointed out that baleen had a pattern of ridges and hollows along the plates, and suggested that they might be like the growth rings on the horn of an ox, which could reveal the age of the animal. Modern biologists, however, ignored this and relied on the size distribution, to the possible detriment of the whales and the people of the whale.

The need to respect the Eskimos' primal rights almost certainly delayed the votes that signal to the public the end of commercial whaling, if not the end itself. It has raised the horrible prospect of a third category of whaling, neither commercial nor aboriginal; Norway and Japan both maintain that their whalers are as dependent on whaling as the Eskimos are on Bowheads, which is solipsistic nonsense. And in the end the Bowhead, reduced by commercial depredations, may not be able to sustain even the limited Eskimo kill. In which case, will all the horse-trading have been worth it?

I do not normally subscribe to the conspiracy theories so popular with some journalists. They are able to see hidden links between the most unlikely events, and frequently ascribe unfathomable machinations to the secret workings of what they generally refer to as the agro-military business complex. I am usually unable to follow their reasoning, but in contemplating the Bowhead saga I do sometimes wonder if the whole thing hasn't been part of a giant conspiracy. About 40 per cent of the United States' remaining oil and gas reserves lie beneath the frozen waters of the Beaufort Sea. At the moment the presence in those waters of an endangered species, such as the Bowhead whale, acts as some sort of brake on wanton exploitation. If the Bowheads were to vanish, things would be a lot easier for the resources companies. And if the Bowheads disappeared so, probably, would the troublesome demands of Alaskan Eskimos. The North Slope would be opened up for another oil rush. Perhaps that is just too cynical, but the evidence suggests otherwise.

Canadian oil companies are already shipping oil across the Arctic for storage in artificial islands. Giant Japanese super-tankers then moor alongside the island to load up for the journey home. At the moment the Canadians say that this is just a technical exercise, to see whether it can be done, but the truth is that Japan is desperate to assure itself of a supply of oil and gas, and Canada is eager to be the supplier.

The US Department of the Interior continues to open up the land of the North Slope for oil and mineral exploration. In the short Arctic summer of 1987, the 210 inhabitants of Kaktovik, one of the eight whaling villages in Alaska, were invaded by politicians, congressional staff members, high-powered oil company executives and, snapping at their heels, journalists and environmentalists. Kaktovik catches Bowheads in the autumn, as they turn and head south after the summer feeding binge. But they may soon have to contend with drilling rigs behind the village and out at sea, for the oil men want to move in on what they see as the last unexploited reservoir in the United States.

Conservationists would like to see more energy conservation. The Reagan administration insists that it needs to know where the oil is, not to use it but to avoid being held hostage by foreign suppliers. The oil men say that Prudhoe Bay, the biggest Arctic field, will start to run out in 1990, and they are looking to Kaktovik not to check the reserves but to find oil to pump down the Alaska pipeline. And the Bowhead whales are still in the way.

In a narrow sense, the Alaskan Eskimos stand to profit from all the oil activity. It will bring them money, just as previous rushes to the Arctic have done. But lately they have said that their culture is more important to them than any cash income. That was why they effectively held the US government to ransom over the Bowheads in the first place. Now, there is a good chance that the foundations of their culture will be severely eroded. If the noise and disturbance of the oil works fail to affect the Bowheads, there is every likelihood that oil will one day spill onto the water. When it does, it will clog up the Bowhead's baleen plates and, unable to sieve a living from the sea, the whales will starve to death.

Back in 1980, when the block quota was first agreed, an Alaskan journalist took me aside to give me what I hoped was an insider's perspective. "It's a suicide pact," he told me. "An endangered culture finishing off an endangered species." I hope he is proved wrong.

Sleazy Deals

You may charge me with murder—or want of sense—
(We are all of us weak at times):
But the slightest approach to a false pretence
Was never among my crimes!

Fit the Fourth: *The Hunting*

"Sleazy deals" is not my phrase. A member of the United States' delegation to the IWC used it to describe for me the bargaining that his commissioner had conducted to assure a three-year quota for the Bowheads in 1980. It is an apt form of words, and ought to have occasioned no surprise. But it is surprising to discover how often the spirit of the International Convention for the Regulation of Whaling, 1946, has been undermined by all manner of subterfuge. Quite apart from objections—legitimate perhaps but unfortunate—the whaling nations have used a plethora of techniques to assure themselves of a continued take.

Some sound like far-fetched stories of international intrigue and heavy-handed persuasion, and are quite hard to support with evidence.[1] There is the matter of Japan and the Panamanian sugar deal, for example. In 1978 the Panamanian delegation, under the instigation of Jean-Paul Fortom-Gouin, was one proposer of a moratorium. Fortom-Gouin was removed from the delegation at the last minute and the proposal was withdrawn, amid rumours that the Japanese had threatened to renege on a deal worth $9.75 million to finance a new sugar refinery in Panama. Japan also threatened to rescind a contract to buy 50,000 tonnes of sugar.

That same year Japan had imported 6,030 tonnes of whale meat from countries not affiliated to the IWC and therefore not hunting within the quotas. This included 2,776 tonnes from Cyprus, 2,645 tonnes from Spain and 597 tonnes from Somalia,

---

[1] Unless other details are given, allegations in this chapter come from ECO, Carter and Thornton (1985), and various other sources.

all the proceeds of pirate whaling operations. The figure had been considerably higher, but officials omitted the imports from South Korea and Peru retrospectively after those nations had joined the IWC. Whaling outside the IWC could not but undermine the Convention, and would have made no financial sense without a ready Japanese market. The Japanese, however, consistently fudged the question of where their whale meat was coming from. It was only after Greenpeace published the booklet *Outlaw Whalers*—which exposed the often tortuous connections between Japan and the pirates—that Japan reluctantly changed its position. It stopped buying from pirates and made sure that its other suppliers joined the IWC.

It was rumoured that Japan paid the fees needed for those countries to become IWC members. In 1983, when Peru was two years behind with its dues, the manager of the Victoria del Mar whaling station at Paita (controlled by the Kinkai Whaling Company of Japan) gave the Peruvian Fisheries Minister a cheque for $63,000 made payable to the International Whaling Commission. This would have been enough to pay the outstanding amount of £30,000,[2] but at the last minute the Ministry of Foreign Affairs got cold feet and declined to use the money. Peru should by rights have had its voting privileges removed for being in arrears, but the IWC decided to suspend its procedural rules to allow Peru a continued say.

The National Fisheries Society in Peru, a "laundry" for Japanese whaling money, also paid for a delegate with invalid credentials to replace the accredited official delegate at the meeting in Botswana of the Convention for Trade in Endangered Species (CITES). Peru was originally to have abstained on all issues. The fake delegate was instructed to vote against all conservationist measures, especially those relating to whales.

Overall, Peru has received considerable largesse from Japan. After the ravages of El Niño, the perturbation of ocean currents that brought disaster to the west coast of the Americas, the Japanese gave Peru $500,000 to help repair the damage. The grant was handed over on the eve of the 1983 IWC meeting, some three months after the worst of El Niño's storms. Peru maintained its objection to the moratorium.

---

[2] The IWC Secretariat, being based in England, operates in sterling.

Peru is not the only country to be linked with Japan. When the Seychelles joined the IWC in 1979 it too was reported to have received unrefusable offers from the Japanese. The story was that unless the Seychelles was more accommodating, it would not receive any development aid from Japan. Oddly, in the three years between independence and joining the IWC, the Seychelles had not received a single yen of aid from Japan. Parties of Seychellois ministers were flown to Japan for high level talks; the visitors were quoted as saying "we retract our demand for a ban on whaling and ask other nations not to oppose whaling", a change of position that the Japanese press interpreted as stemming from the realisation that "it is not in their best interests to kick Japan around". Advisers to the Seychelles tried to pour cold water on the stories. Nevertheless, the rumours were so strong, with stories of Seychellois parliamentarians and their families enjoying all-expenses trips to Japan, that the government was eventually forced to issue a denial.

St Lucia, another strongly conservationist country, has also been subjected to repeated hassles from Japan. In 1982 St Lucia was working hard for the moratorium. Alan MacNow, on behalf of the Japanese, attempted to discredit Peter Josie, the St Lucia commissioner, by describing Francisco Palacio, Josie's alternate, as "an impostor, a terrorist and a mercenary". MacNow is a fixture of IWC meetings, retained by the Japan Whaling Association as a public relations consultant. The government of St Lucia reaffirmed its support for Dr Palacio, and MacNow was made to look a fool at the press conference he called to take credit for "unmasking" Palacio. The Japanese commissioner apologised for the incident. He unfortunately spoiled this genteel performance by claiming that there was no link between MacNow and the Japan Whaling Authority. MacNow was registered with the US Congress as a lobbyist for the Japan Whaling Authority, at a fee of about $400,000 a year.

Palacio came in for further trouble the following year, perhaps as a result of his personal intervention with the Peruvian President, which was probably of considerable influence in persuading Peru finally to waive its objection to the moratorium. Palacio conducts his academic work from the University of Miami, which received documents seeking his dismissal for alleged professional incompetence. The government of St

Lucia and other IWC countries also received similar packages of documents, which were apparently sent by Japanese whalers or their supporters. The allegations proved baseless, and Palacio again received support from several individuals and governments. The University of Miami has not seen fit to dismiss him.

St Lucia, however, has now got rid of Palacio, who is no longer even a scientific adviser to the delegation. St Lucia did some other very strange things before the 1986 meeting: it quit the Scientific Committee, withdrew an agenda item on outlaw whaling, gave a warm welcome to Congressman James Dymally and Alan MacNow, both staunchly pro-Japanese, and withdrew an earlier offer to host the 1987 annual IWC meeting. The prime ministers of St Lucia and St Vincent and the Grenadines had both been invited to Japan in 1985. Each was asked to change his policy on whaling, and asked also what he wanted in return. According to one conservationist, who explained the deal to me on the eve of the 1986 meeting, "the PMs had a list, and we could well see a massive Japanese aid plan in the eastern Caribbean. As a price, Palacio had to go, but he has achieved a great development boost for the region. And it's probably too late for Japan to get any real benefit in this meeting."

Too late it may have been, but at the Malmö meeting the Japanese continued to pursue opponents of commercial whaling. On the afternoon of the penultimate day they challenged the credentials of Roger Payne. Payne, who discovered the song of the Humpback, is currently scientific adviser to the government of Antigua and Barbuda, another of the little Caribbean island states. Antigua is in arrears over its IWC dues, so does not have a vote in the plenary sessions. Japan, nevertheless, wanted Payne out, saying that his government had not sent his accreditation to the IWC. Payne refused to go. The chairman, Ian Stewart of New Zealand, hates any thought of controversy, and after politely asking Payne to leave, and receiving a resolute refusal, adjourned the plenary session. Through the glass doors Stewart could be seen huddled with Payne, Sir Peter Scott of the World Wildlife Fund, and various other conservationists. What was said is unknown, but when the plenary resumed, Payne was no longer an adviser to Antigua and Barbuda; he had become a non-governmental observer attached to the Seal Working Group.

174

After the incident Payne contacted his commissioner, who immediately telexed the IWC to "reiterate that Dr Roger Payne is acting as scientific adviser to the delegation of Antigua and Barbuda". Chairman Stewart ruled that the telex "doesn't quite comply" with IWC requirements. After a deliberative pause he continued, "Because telexes have some uncertainty attached to them they require confirmation in writing," and hence his ruling that Payne be removed from the delegation stood. One wonders what the Japanese sought to achieve by this particular attack. Why had they waited until one day from the meeting's end to launch it, if not to make rebuttal impossible? As Payne noted, "This is not the first occasion on which delegates from small countries have been challenged in an apparent attempt to intimidate them."[3] He also said, "I feel an apology is in order from those who precipitated this unpleasant episode." None was forthcoming.

The horse-trading over individual whale stocks has also been fierce. (Some has already been detailed in other chapters.) In general it has been noticeable that countries have been rewarded with whales for staying within the IWC and lending support, or at least withdrawing opposition, on selected issues.

There have been glorious debates within the IWC over just what constitutes a whale. The Convention applies only to those whales specifically mentioned in it. Other species, often as hard pressed, gain no protection. In 1980, for example, the USSR factory fleet announced that it had taken almost 1,000 killer whales from the Southern Ocean. This was a violation of the ban on factory fleets, but as killer whales were not mentioned in the schedule it was not really wrong. Killer whales were then added to the list, and became part of the involved discussion about the role of the IWC in regulating so-called small Cetaceans.

Bound up in this debate are questions of rights over 200-mile exclusive economic zones. Indeed, small Cetaceans seemed at times like a code word for national jurisdiction, the factor believed to have been responsible for Canada quitting the IWC in 1981; Canadian Eskimos take narwhals and beluga whales in the Arctic, and Canada was not about to allow the

[3] Statement by Scientific Adviser to Antigua and Barbuda, 13 June, 1986.

175

IWC to have any say over these small Cetaceans in Canadian waters. The fact that narwhals and belugas are not covered by the Convention did not seem to matter.

The whole question of just what constitutes a whale, within the meaning of the act, produced some laughable conclusions. Belugas and narwhals, it seems, are not whales. They are dolphins. The Sperm whale, which by that reckoning is also a dolphin, is, however, a whale. Small Cetaceans, such as the Bottlenose whales and beaked whales (both dolphins, if anything), are often bigger than acknowledged great whales like the Minke.

It is all very confusing, but the prize for best semantic wrangle goes to the Danes. They object to the IWC having anything to do with any of the smaller whales, a position that seems to stem from Denmark's relationship with Greenland and the Faroe Islands. Hunting of Pilot whales and belugas is a local tradition there, but the Danish reason for keeping the IWC out of these affairs is wonderfully specious. It is that the species in question were not included in a list of the vernacular names of whales drawn up in various languages at the original IWC conference in 1946. In other words, because no one thought to include the English, Dutch and French names of these various species in the Convention at the time, they are not whales at all.

Quite another semantic confusion surrounds the issue of scientific whaling, now the biggest threat to whales. Article VIII of the Convention allows any nation to issue itself with a permit to take any number of whales "for purposes of scientific research". It states further that "scientific" whales "shall so far as practicable be processed and the proceeds shall be dealt with in accordance with directions issued by the Government by which the permit was granted". Add to this the further facts that the number and type of whales to be killed for research are up to the issuing government alone, and are exempt from the quotas, and you may have an inkling of why I am treating scientific whaling as a sleazy issue.

In 1985 the governments of Iceland and South Korea each announced that during the moratorium they would be taking a few whales for science. Iceland intended working on eighty Fin whales, eighty Minke whales and forty Sei whales each year between 1986 and 1989. The proposal also mentioned the possibility of killing an unspecified number of Blue and Humpback whales in the latter half of the project. Korea was less specific, saying only that it planned to take 200 Minke whales a year over the four years from 1986. (Brazil also circulated a proposal for scientific whaling but later, to that country's credit, withdrew it.)

The proposals claimed that the research would provide valuable data. Scientists at the IWC in 1985 were almost unanimous in saying that they neither wanted nor, more importantly, needed those data. The meat would in any case be sold to Japan, and the proceeds used, in the Icelandic case certainly, to maintain the whaling operation in good repair. Conservationists were in uproar. Sir Peter Scott, speaking for most of them, described the proposals as "blatant subterfuge" and a "flagrant abuse of the Commission's procedures".[4] It sounded from the noise almost as if Iceland and Korea had discovered some new loophole through which to drag whales, but scientific whaling has a long and dishonourable tradition; the latest proposals might almost have been predicted.

Scientific whaling went on sporadically in the first decade of the IWC, but generally the numbers involved were below thirty. The USSR at one stage tried to use the permit to extend the season, for fear that whales killed for science during the season might be counted against its quota of Blue whale units. Norway questioned a British proposal to use twelve whales to test an electric harpoon, saying that this was not within the "ambit" of Article VIII; a subsequent attempt to define "scientific research" failed because the committee felt it would be anti-scientific to place any kinds of limits on research. It was in 1962, however, that scientific whaling really came into its own.

The number of whales mentioned in research permits that year leaped to 100, and it was also the first year that the whalers failed to reach their quotas in the Antarctic. In 1963, when the quotas were reduced by a third, more than 500 whales were taken under scientific permit, and although the number of scientific whales declined in later years they stayed well above 100 through the 1960s. Most of these scientific whales were Sperm whales, at that time forming an increasing proportion of

---

[4] Statement by Sir Peter Scott for WWF Press Conference on IWC, 10 July 1985.

the catch. There were many reasons for the interest in Sperm whales. Baleen whales were becoming very scarce, and in any case Sperm oil fetched a higher price than ordinary whale oil. The IWC did not set quotas on Sperm whales until 1970, and during the 1960s they were not included in the Blue whale unit.

The Sperm whales were pursued especially by Japan and the USSR, which had easy access to the Sperm whale grounds of the North Pacific. South Africa, Australia and New Zealand —southern hemisphere countries that had been heavily dependent on coastal whaling for Humpbacks—suffered the loss most as the Humpbacks vanished. It was those countries that used scientific permits to go after Sperm whales, and were between them responsible for the sharp increase in special scientific permits between 1962 and 1963. Of the three, New Zealand was the one that most blatantly used scientific whaling to boost its whaling industry, which nevertheless collapsed after two years.

At the time the scientists were both pleased and apprehensive about the burgeoning of research whaling. They would, it is true, be getting more data, but often those data were of suspect usefulness. And even the scientists were aware that research was being used to prop up the industry. They argued that the commission should have to approve scientific permits before the contracting government could issue them, which might ensure that the research was useful and needed, but the recommendation got no further.

In the mid 1960s the focus of scientific whaling shifted from the southern hemisphere whalers to North America. Gray whales had been almost wiped out towards the end of the nineteenth century by the coastal operations based in California. By the 1960s it was apparent that numbers had increased considerably and the Scientific Committee was curious to learn more about their recovery. In 1966 the committee encouraged the United States to take some Gray whales for science. The Del Monte and Golden Gate fishing companies killed 290 Gray whales under scientific permit between 1966 and 1969.

The new Californian whalers had been going after Humpbacks since the 1950s, but suffered a series of collapses that mirrored the industry as a whole. In 1956 they took 199 Humpbacks; by 1965 the figure was only four. Fins filled the void: 113 in 1965, forty-two in 1966, and five in 1970. Sei whales followed the same trend. Against this background of successive collapses the average take of 72 Gray whales a year for scientific research kept the industry going. The United States withdrew the scientific permits for Grays after 1969. In 1971 the Californian whaling industry folded.

A similar story unfolded at the opposite corner of the continent, off Nova Scotia and Newfoundland. Norwegian and Japanese interests established themselves there in the early 1960s to exploit stocks that had apparently recovered from the very severe depredations of the 1930s. They went after Fin whales, but catches declined steadily and the whalers found it very difficult to meet even reduced quotas. They had invested considerable amounts in the processing stations on land, and in what turned out to be the final two years of the industry the Canadians issued a permit to let them take seventy Humpbacks, which were fully protected, and eighty Fin whales. The industry collapsed anyway.

In the wake of the Stockholm conference of 1972 the IWC decided to set up an International Decade of Cetacean Research (IDCR), starting in 1974. A committee of scientists drew up a research programme which included sampling animals from stocks that had not been exploited, especially the Sei and Bryde's whales of tropical waters. These had been protected by the long-standing ban on factory fleets between 40 degrees S and 20 degrees N, and the proposal called for quite extensive biological sampling to learn more about these populations. Such information could have been very valuable.

Perhaps the most notorious feature of the IDCR proposal was the suggestion that the IWC hire the infamous pirate whaler *Sierra* to conduct the research. The *Sierra* had been operating as a pirate since the 1960s, killing as many whales, including protected species, as it could find. It had done so under various flags of convenience, making a healthy profit by supplying Japanese buyers. The IDCR proposal was clearly an attempt to legitimise the *Sierra* and bring it under the IWC's purview. In the end she never was hired, and continued making inroads on whales in the North Atlantic until sunk by *Sea Shepherd*'s explosives in Lisbon.

Not much came of the IDCR proposals, except for the suggestion that more research be conducted on Bryde's whales. The Japanese took this suggestion to heart; with rorqual whaling all but shut down in the Antarctic they could rescue their industry if their research on Bryde's whales in tropical waters showed they could be exploited. Over the course of the Japanese project 459 Bryde's whales were "sampled", but the most important result to emerge from this research was 3,000 tonnes of frozen Bryde's whale meat. There was a constant problem obtaining a random sample of the population, crucial for any scientific assessment. The whalers preferred to go for the bigger specimens, even when instructed to the contrary. This bias makes any scientific study of the population, the ostensible purpose of the hunt, impossible, although it does of course make sound commercial sense. The whalers did not even carry out the research required of them on those whales that they did catch: of 225 Bryde's whales caught under special permit one year, only forty were measured even though all were supposed to be. The final comment on the scientific utility of the special permits for Bryde's whales is that when the Scientific Committee came to assess the stocks they used data from sightings cruises, in which whales are seen but not hurt, and not information from the sampling.

In the light of these abuses of Article VIII the IWC finally changed the rules slightly in 1979. It added a paragraph to the Schedule saying that research projects should be submitted to the Scientific Committee for comment and review, and set out the minimum requirements for the information needed to assess special permits. There was opposition to this change, but it did indeed pass into the Schedule, although it still allows each country the final say, as set out in the Convention. Between 1979 and the 1982 vote on the moratorium the Scientific Committee received three proposals for research.

The Peruvians wanted to catch two or three Blue whales. This was to establish whether the Blue whales off the coast of Peru were indeed normal Blue whales, or whether they might be the slightly mysterious Pygmy Blue whales. The Peruvians also wanted skeletons for Peruvian and Ecuadorean museums of natural history. The Scientific Committee expressed its reservations, and museums in Canada and California offered spare skeletons, so that in the end the Peruvians did not go ahead.

Chile proposed taking 100 Sei whales in 1981. This came hard on the heels of a Japanese sighting cruise that had reported substantial quantities of Sei whales in the waters off Chile. The Chileans had already taken large numbers of Sei whales with the help and guidance of Japan, and had not properly examined the data from those catches. They had also been exposed for using an illegal factory operation, the *Juan 9*, financed by Japan. The scientists suggested that if the Chileans wanted to contribute to the corpus of knowledge on whales they would do better to turn their attention to compiling and analysing earlier data, rather than collecting new specimens. The Chileans, however, did continue to take Sei whales after they had been protected.

The final proposal before the vote on the moratorium came from the Faroe Islands. The Fin whales around the islands have been protected since 1976, but a hunt has been carried on, with the full knowledge of the Danish government, since that time. One man, Herman Hermansson, conducted this operation, which was explained to the IWC year after year as a "breakdown in communications" between the Danish government and the Faroese fishermen. In 1981 the Faroese Home Rule Office wanted to ask the IWC directly for a quota of ten Fin whales, but the Danes judged that this would be badly received. Instead Hermansson drew up a plan with the Fisheries Investigations office. His company would conduct research on almost exactly the same number of whales each year. The Scientific Committee did not think that nine whales a year would add much to its knowledge of the stocks and advised against the proposal, but the Danes went ahead and issued an everlasting special permit anyway.

The scientific information provided by each whale the Faroese kill is laughable—date, sex, length and in some cases stomach content. Each year the Faroese have said that no more Fin whales would be killed. Each year they have been. Even the Director of the Fisheries Investigation Office recognises the operation for the sham that it is. The Danish newspaper *Ekstra Bladet* reports him saying "the research is not the most important part of the whaling. What is most important is the whale meat, as a source of food and income." Very little of the meat is eaten locally. Most is sold to Japan and provides income for Mr Hermansson.

1985, the year in which the moratorium was to start, saw the Icelandic and Korean proposals. They had been foreshadowed by reports in the Japanese press that the Whaling Council, which advises Japan's Fisheries Agency, had recommended that Japan abandon commercial whaling in the Antarctic and switch instead to research whaling. The meat would be sold on the domestic market, as usual, and the fleet in the first year of scientific whaling would be identical with that used in the last year of commercial whaling. This notion was shelved after Japan reached its "understanding" with the United States (see p 190), but it remains a prominent option. In fact on 5 April, 1985, Moriyoshi Sato, Japanese Minister of Agriculture, Forestry and Fisheries, said that "the government will do its utmost to find out ways to maintain the nation's whaling in the form of research or other forms", a threat that has been repeated many times since. In 1987 Japan launched its proposal to kill 875 whales a year for at least twelve years.

Special permits to take whales for science have been attacked from two different directions: on the one side are the genuine scientists, who do not see how the proposed research will shed any light on the unsolved problems of whale biology; on the other are the conservationists who have been trying to remove the financial support for scientific whaling. Neither seems to be making much headway.

In the past fifty years more than half a million whale carcasses have been available for scientific scrutiny, and still the enigmas remain. Sidney Holt speaks for most whale researchers when he says, "looking at the entrails of more dead whales will not help". In fact, much of the information collected by commercial whalers remains unanalysed, because the whaling nations have not been very forthcoming with it. This is one of the major complaints that the new group of four scientists —Doug Chapman, Sidney Holt, Bill de la Mare, and Roger Payne—have against the Norwegian whalers.

The argument is about how to use poor data. The Norwegians admit that the data are so bad they reveal nothing about the current state of the stocks. The group of four say that despite their inadequacies the data point to a serious depletion in the stocks, perhaps to less than a tenth of the original size. Either way, the conclusion ought to be the same. If the data are useless there ought to be a pause while data that unequivocally support whaling are collected, research that requires no killing. And if the data are accurate, the stock fully deserves the protection the IWC voted it. But there is a more serious problem.

Norwegian scientists said, after the vote to protect the Minkes, that not all catches had been reported as required to the International Bureau of Whaling Statistics. That is why it appeared to the Scientific Committee that the whalers were finding it harder to catch each whale, because they actually caught more whales than they originally owned up to. To which the scientists have a simple reply: they cannot manage the industry if the whalers will not tell the truth about their take. The group of four goes further. If the Norwegians are lying, and if their lies provide the basis for protecting the whales, Chapman and his colleagues believe that "natural justice" dictates that the data be taken at face value and the Norwegian industry be shut down.

The scientific arguments against scientific whaling did not have much impact on the whalers. Other conservationists have tried an economic approach. Resolutions were put before the IWC concerning the products derived from any whales taken for science. Sweden and Switzerland had condemned the Korean and Icelandic proposals when they were first announced in 1985. At Malmö they put forward a resolution forbidding international trade in meat from research whales. That could never have been adopted; Iceland threatened that if it were there would be "severe repercussions in the attitude of Iceland towards the Commission". Oman proposed a compromise. Meat from research whales should all be consumed locally. Iceland could not swallow that either; there are not enough Icelanders to eat all the whales they plan to kill for science. A final proposal from the compromise-hungry chairman suggested that scientific whale meat be "primarily for local consumption". This was adopted by consensus.

Although weak, the resolution on scientific whales did at least indicate that the IWC as a body is aware that research whaling is often simply a front for commercial whaling. It might have deterred the Japanese from supporting too heavily scientific whaling by other countries. That, in turn, might have dampened those countries' enthusiasm for research whaling.

It did not, however, stop the Norwegians announcing, about three weeks after the end of the IWC meeting, that, like Iceland and Korea, they would abandon commercial whaling during the moratorium but, again like Iceland and Korea, they would continue to whale for scientific reasons. They had earlier said that they would voluntarily reduce their catch from the north-eastern Atlantic stock of Minke whales, which is the subject of so much argument among the scientists, to 400 whales in 1986. Despite a two-week extension of the season, Norwegian whalers managed to take only 379 of the 400 Minkes they had hoped for.

After toying with the idea of scientific whaling at the 1986 meeting in Sweden, when the IWC returned to Bournemouth in 1987 scientific whaling was the most important item on the agenda. Japan and Norway had announced the details of their plans to stop commercial whaling and start scientific whaling. Iceland and Korea had been at it for a year. And the conservationist countries came prepared to do battle on behalf of whales killed for science.

The Scientific Committee of the IWC had already been charged with deciding whether any nation's plans to kill whales under a special permit met certain criteria for good science. The Americans put forward a new proposal, the essence of which was to ask the IWC itself to notify member governments if their research plans did not come up scratch. The resolution also recommended governments not to allow any scientific whaling if the IWC had decided the scientific criteria were not met. The effect of this would be to help the United States enforce its sanctions, if it chose to do so.

The conservationist attack in 1987 focussed squarely on the science in scientific whaling. Sir Peter Scott publicly returned the Order of the Falcon conferred on him by the government of Iceland, and he decried scientific whaling as a "flagrant breach" of the spirit of the Convention. The whalers themselves replied that it was their duty to continue whaling for science, to provide data needed for the comprehensive assessment that is supposed to be going on during the moratorium. The members of the Scientific Committee, however, decided to judge research whaling, both that already carried out and that planned for the future, on its scientific merits. Their conclusion seemed to be that it is at best useless, and at worst may be damaging the very stocks of whales it is supposed to assess.

Korea took sixty-nine Minke whales, fifty-two males and seventeen females, from the Sea of Japan in the 1986 season. In presenting the results of his country's research programme Dr

Yeong Gong got a thorough drubbing from the Scientific Committee. Why, the scientists wanted to know, had no ovaries been taken from the females, even though the testes of the males had been weighed? Ovaries could provide data on pregnancy rates. No analyses were made of stomach contents, and very few earplugs—useful for estimating the age of the whales—were taken. The Koreans did undertake some non-lethal research, a sightings cruise on which they looked for whales in an effort to estimate how many there were. Alas, although they saw three whales, the Korean whale scientists did not note how far they had travelled on the cruise, so it proved impossible to use their sightings to estimate the abundance of whales.

All in all, the Scientific Committee was dismayed by the Korean research. "The information cannot help with any significant management question, nor is it of any significant scientific value, as only a fraction of the potentially available biological data were collected," the committee's report noted. The Korean proposal fell far short of the guidelines for good research, and "may have caused further reduction of this depleted stock, or at best inhibited its recovery". As to the Korean plans for future whaling seasons, the Scientific Committee did not think that the science in Korean scientific whaling would improve. The proposal provided "no new information", and there was no chance that future plans, which are no different, would be any better.

Iceland fared only slightly better. It took seventy-six Fin whales and forty Sei whales in 1986, and managed to weigh twenty-two of the Fins and twenty of the Seis. The Icelanders also obtained some blood samples and hormone levels, but, Sidney Holt said, "They have little relevance to what we're trying to do." Again, the Scientific Committee did not feel it could offer Iceland's scientific whaling unqualified support.

Norway has not yet conducted any scientific whaling. Having objected to the moratorium, Norway is free to continue whaling, but the Fisheries Department did announce that it would stop commercial whaling in 1987. Scientific whaling will start with the 1987/88 season. In order to meet the objections of scientists and conservationists, Norway commissioned an independent review of the state of the stock of Minke whales in the northeast Atlantic. The report, by two Norwegian and two

British scientists, not members of any lobby on whaling, suggests that some useful data could be collected by killing whales, although just what they mean by "useful" is not spelled out. But it also pointed out that the research programmes proposed by Norway and Iceland "lack a sufficiently clear definition of purposes and priorities". Norway has not yet notified the IWC formally of its proposed special permit, so the Scientific Committee was unable to comment on it.

The final nation to discover a sudden interest in whale research was Japan, whose proposals are far more detailed and require far more subjects than the other scientific whalers. In essence, Japan plans to take 825 Minke whales and fifty Sperm whales from the Antarctic each year, more or less forever. The Minke whales are to be used in an entirely new way, to estimate simultaneously how fast the population is growing and how many could be harvested. Sidney Holt says the Japanese research "is nonsense". Dr Syoiti Tanaka, author of the research plan, told me in Bournemouth that he "believes it will work".

Tanaka recently retired as professor of fisheries science at Tokyo University, and is Japan's foremost fisheries statistician. His plan calls for Japanese scientific whalers to steam to points determined at random, and then to capture the first whale they see. In this way they catch 825 whales. Four years later, they come back and catch another 825 whales. (The proposal calls for each of four areas to be sampled in rotation, so the whalers are kept busy every year.) Each sample will provide the scientists with an estimate of the age distribution of the whales, that is the proportion of the population that falls into classes of different age.

In the research proposal, Tanaka claimed he would be able to use these two age distributions to work out both the natural mortality, that is the chance of a whale in any given age class surviving to the next older class, and the recruitment rate, that is the number of whales entering the youngest age class. He would do this using a mathematical technique called iteration, which works like this. First you guess at a value for, say, the mortality. You apply it to the data you have collected, and that produces a more accurate estimate. You apply that again, get a more accurate estimate still, and so on. Eventually you converge on the correct value, or a figure very close to it.

The difficulty, as Holt and other scientists tried to explain to the Japanese, was that Tanaka intended to do two sets of iterations with two variables, mortality and recruitment rate. That, they said, simply cannot be done. Tanaka's method "will converge on a solution", said Bill de la Mare, the Australian population biologist, "but the trouble is it isn't a unique solution".

To see what de la Mare means, imagine you have the equation $A + B = 10$. You can guess that A is 5, in which case B will also equal 5. That is a solution. But it is not a unique solution. A could be 7, making B equal to 3. The same goes for trying to estimate both natural mortality and recruitment from snapshots of the population. You must have a second equation. If $A - B = 4$, there is only one solution, A is 7 and B is 3. If you do a sightings cruise, you could use the estimated number of whales to help derive the other figures, but you cannot do it from hunting alone.

Of course this is a simple example, but it is an exact analogy of Tanaka's proposed method. Doug Chapman, an American biologist, described the method as "not consistent with known theory". Tanaka said "I am sure it converges." Kees Lankester, a mathematically inclined biologist with the Dutch delegation, was adamant in his criticism. "You can't do it. We know you can't do it. We don't know why he doesn't know you can't do it." There seems to be a genuine failure of communication between the Japanese and most of the other biologists, but an easy way exists to settle the argument.

The whale biologists already have an acceptable computer model of a whale population, on which they assess various management regimes (see "Science", p 140). Tanaka could try his method on simulated data, and if he came up with the correct results, the ones that have been programmed into the computer model, he would know he had been right all along. Bill de la Mare, who runs the computer model, thinks that the whole matter could be settled with a few weeks' work, but Tanaka sees no need. He is confident that his technique will work. The Japanese government agrees, to the tune of a subsidy of $4 million for the first expedition.

The method for estimating the parameters was at the heart of the disagreement in the Scientific Committee, but there were other problems with the Japanese research plan, too. The "random" catch is supposed to overcome the biases inherent in previous data from whales hunted for commerce. But the new data will also be biased, because if the whalers take the first whale they see they are likely to take solitary whales and large whales more often than they "should". Within the circumlocutions of the Scientific Committee's report, Japan's proposal, like Korea's and Iceland's, would not produce good science. The whalers may not choose to call it commercial whaling, but they should not call it scientific either.

The scientists were clear in their condemnation of the quality of the various research proposals, but that did not mean that the whalers were going to submit to scientific judgement. All manner of arguments were produced, including a very lengthy legal analysis from Iceland, which ended with a threat to take those nations that supported the American resolution to the International Court of Justice in The Hague. In the end, though, the proposal was accepted by nineteen votes to six, with seven abstentions.

Three nations now sprang to the attack. Each had come prepared with yet another resolution, acting on the provisions of the newly adopted American resolution. They asked the IWC to notify the governments concerned of the deficiencies in their plans for scientific whaling, and recommend those governments not to do it. The United States picked on the Republic of Korea, Australia had chosen Iceland, and the United Kingdom's target was Japan. After a night of urgent consultations with home governments, the resolutions were put to the vote, and all were adopted. The resolutions passed at the 1987 meeting may now give conservationists in the United States a better case for imposing sanctions, which in turn may make the die-hard whalers think again.

The moratorium is due to be reassessed in 1990. Scientific whaling must be seen against this background. It might not be as profitable as straightforward commercial whaling, although it is hard to see why not. But it could provide funds to maintain the whaling fleets at least until 1989. If whaling has been going on during the pause, ostensibly to provide information about the effects of the pause on whaling, there will have been no pause.

There is uncertainty about whaling, but it is unlikely to be resolved by further killing. Even if it were, scientific whaling need not be funded by the proceeds from the sales of scientific whales. Large industrial corporations fund their research from profits on other endeavours. They do not, for example, sell drugs that have not yet been tested in order to pay for the testing of those drugs. And yet that is what the whalers now say they must do. If they are so sure that their research will show that the stocks are capable of exploitation, let them have the confidence to invest in that research.

Scientific whaling is not whaling for science; it is a way around the will of the International Whaling Commission. The truth is that the IWC is sadly lacking in teeth of any kind. Nations have always been free to object, just as they have always been free to ignore the IWC completely and leave. Scientific whaling is just another expression of that freedom. The only curbs that keep IWC countries within bounds are the harsher realities of international politics and trade; if the whalers could do their business undercover, I am sure they would. As it is, when they are exposed to publicity some of them at least change their ways.

The single real force behind well-intentioned decisions of the IWC is the United States government. It has two amendments to fisheries acts, both of which are supposed to swing into play as soon as the US Commerce Department certifies that a nation is undermining the effectiveness of the IWC. The Pelly amendment allows the President of the US to halt fisheries imports from the country concerned. The Packwood/Magnuson amendment calls on the Commerce Department first to halve, and then to cut completely, the offending country's allowance of fish from within the exclusive economic zone of the US. Those amendments were enacted specifically to provide conservationists with the muscle that international treaties conspicuously lack.

They have been used. In 1985 the US penalised the USSR for taking too many Minke whales the previous season. The Packwood/Magnuson amendment was employed to halve the Soviet Union's quota of Alaskan fish. The USSR had only recently resumed fishing in Alaskan waters; its allocation was small and the cost of the sanctions is unknown, but probably insignificant compared to the extra Minke whales taken.

Japan is quite another matter. US sanctions would hit

Japan, which takes fish worth almost twenty times its whaling industry from Alaskan waters each year, particularly hard. But while the Soviets have been hit, the Japanese have not. The Commerce Department decided all on its own not to use the only stick it has to enforce decisions of the IWC. And the lucky recipient of this generosity is the only country that really matters.

The ban on Sperm whaling came into force in the autumn of 1984. The Sperm whale season in Japan normally starts on 1 October. The start was delayed while Japanese and American negotiators huddled together. The catcher boats eventually set out from the northern port of Ayukawa on 16 and 17 October, but they were recalled within a week, apparently in fear of the US sanctions. At first it seemed that the Japanese were adhering to the ban, despite their official objection to it. In fact they were involved in long discussions with the United States government to see whether the whaling industry could salvage something from the wreckage.

The catchers left port again. On 11 November they returned to land two Sperm whales at the port of Wadawura, within their rights under the Convention but nevertheless contravening the ban. That act should have laid the Japanese open to certification and the full effect of US sanctions.

It was not at first clear whether the two whales represented an independent act by the catchers, in defiance of their own government, or whether the government had permitted the catch. Had the whalers gone out again of their own accord? The answer came in an announcement from the US Commerce Department, which had reached "an understanding" with Japan, such that if the Japanese promised to take certain specific actions, the Commerce Department would not certify them for undermining the IWC.

The full text of the "understanding", when it emerged, appeared to rob the IWC of any role in the regulation of the hunt. It referred to "catch limits acceptable to the Government of the United States", and made no mention of the IWC.[5] The agreement allowed Japan to take 400 whales in the 1984/5 season. It also said that if Japan were to withdraw its objection by 13 December, 1984 then the US would allow it another 400 Sperm whales in the 85/6 and 86/7 seasons. The Japanese have

[5] Anon. (1984).

190

instead given an undertaking to withdraw their objection on 1
April, 1988, an auspicious date for the end of their understanding with the United States. Another deal applied to the full
moratorium on commercial whaling, and again if Japan withdrew its objection it could keep whaling for two seasons after
the ban came into force. Then, however, the whaling has to
stop. Or so the agreement says.

Conservationists were incensed. A consortium of nine organisations immediately brought legal actions to prevent the Commerce and State Departments from making any arrangement
with Japan and to force them to implement the sanctions. One
possible problem concerned the wording of the legislation. The
Packwood/Magnuson amendment refers directly to countries
that "diminish the effectiveness" of the IWC. Japan had acted
within the constraints of the IWC by filing an objection to the
ban on Sperm whaling. But it had killed Sperm whales, which
the IWC would rather it had not. So was it undermining the
IWC's effectiveness? "There's no simple answer to that problem," was the reply I got from a spokesman for the Commerce
Department. "It's a judgement call."

The judge called it, in favour of the conservationists. In early
March 1985, Charles Richey, a district judge in the District of
Columbia, seemed to have no trouble deciding that the US was
bound to impose the sanctions of the Packwood/Magnuson
amendment. This agreed with Senator Bob Packwood's intention. He had meant for the amendment to be mandatory, he
said, and the decision delighted him and the conservationist
consortium that brought the case. The US government immediately announced its intention to appeal. While all this was
going on, the Japanese caught their 400 Sperm whales.

The appeals procedure took its time—conservationists had
been hoping for a decision before the IWC meeting in July
1985—but eventually, on 6 August, 1985, down came the
judgement. The Secretary of Commerce, decided a three-man
panel of the full Federal Appeals Court, did not have the
authority to enter into any agreement with the Japanese. He
was bound by law to impose the sanctions: "Where a foreign
nation allows its nationals to fish in excess of recommendations
set forth by an international fishery conservation program, it
has per se diminished the effectiveness of that program." In
such cases, said Judge J. Skelly Wright, the imposition of

191

sanctions "is mandatory and nondiscretionary". As, presumably, it had been with the Soviet Union?

Confused by the discrepancy between its treatment of Japan and the USSR, I asked the Commerce Department to clarify the position. The spokesman told me, "It's quite different. The Japanese came to us and asked that we sit down and discuss to see if there was a middle ground so they could phase out whaling without incurring US sanctions. Conversely, we warned the Russians that if they took a larger share [of the Minke whale quota] we intended to invoke the sanctions." The Soviet catch is purchased by the Japanese, who had three years' notice to "phase out whaling" and did nothing about it. Now, the court said, the US Commerce Department was bound to do something about it.

The Commerce Department, faced with unpalatable decisions from three successively higher courts, took its arguments to the Supreme Court, which decided that the case was indeed important enough to merit its attention.

At issue, according to the lawyers, was the independence of the executive and legislative branches of government. The Packwood/Magnuson amendment, as passed, seemed to require the Secretary of the Commerce Department to take certain specific steps. If a country was undermining the effectiveness of the IWC, the Secretary of Commerce had to certify that it was. Having certified the wrong-doer, he was bound to first halve, then eliminate, its allowance of fish from American waters. The Secretary could not ignore those laws, said conservationists. Oh yes he could, the Secretary replied.

Senator Packwood himself had written to Malcolm Baldrige, Secretary of Commerce, some four months before the first agreement between the US and Japan was signed, saying that "it has been assumed ... that a nation which continues whaling after the IWC moratorium takes effect would definitely be certified. I share this assumption since I see no way around the logical conclusion that a nation which ignores the moratorium is diminishing the effectiveness of the IWC." Packwood went on to ask Baldrige for an assurance that certification would be automatic for any countries whaling after the ban. This Baldrige gave him, "since any such whaling . . . would clearly diminish the effectiveness of the IWC". That seems plain enough, but in the event Baldrige changed his mind.

The Supreme Court heard the arguments on 30 April and two months later, on 30 June, 1986, delivered its judgement. "The legislative history does not indicate that the certification standard requires the Secretary, regardless of the circumstances, to certify each and every departure from the ICW's [International Convention for the Regulation of Whaling] whaling Schedules."[6] The agreement was legal after all.

Conservationists were a bit dismayed, but took some comfort from the deep division within the court: the decision had been passed by five votes to four. They also noted that one of the most conservative judges on the Supreme Court bench, Justice Rehnquist, was one of the four who dissented. Justice Marshall, for the dissenters, stated that it was precisely to avoid any discretion on the part of future Secretaries of Commerce that Congress had passed the mandatory sections of the Packwood/ Magnuson amendment. "The Secretary would rewrite the law," Marshall noted, something he could not agree to. The plaintiffs lost the case, but they took heart from the dissenting judges' concluding remarks, finishing as they did with the words of Herman Melville:

> I am troubled that this court is empowering an officer of the Executive Branch, sworn to uphold and defend the laws of the United States, to ignore Congress' pointed response to a question long pondered: "whether Leviathan can long endure so wide a chase, and so remorseless a havoc; whether he must not at last be exterminated from the waters, and the last whale, like the last man, smoke his last pipe, and then himself evaporate in the final puff."

It is hard to evaluate the effectiveness of the US sanctions. Shortly after the Supreme Court decision the Japanese did announce that they will withdraw their objection to the moratorium on 1 April, 1988, provided the US takes no further legal action. Despite the sanctions, the Japanese have continued to whale with no practical regard for the moratorium; would they have taken more whales if there had been no possibility of economic reprisals by the United States? Nobody knows. The US has also used its sanctions to threaten the other countries that continue to "diminish the effectiveness of the IWC".

Norway felt the threat of US sanctions, a blockade against

[6] Supreme Court of the United States, Nos. 85–954 and 85–955, III.

fish imports worth 1.2 billion kroner (£103 million). In response the Foreign Minister announced in July 1986 that Norway would cease commercial whaling in 1987. But it will continue catching whales for science. The quota for 1987 is to be 325 whales, fifty fewer than 1986's kill, which had failed to reach the target level of 400.

Iceland's only whaling firm changed the dates of its 1986 summer holiday to avoid the United States' threat to use the Pelly amendment to embargo imports of fish from Iceland. The whalers began their break on Sunday, 27 July, instead of the more usual date of 10 August, in response to a request from the Icelandic prime minister, Steingrimur Hermannsson, who said he had made the request to facilitate negotiations with the United States.

Iceland had intended taking 120 Fin and Sei whales in 1986, and when the holiday began seventy-six of these had already been caught. The US had said that it would ban imports of fish from Iceland, unless Iceland's whaling came into line with the IWC's regulations. The problem, as far as the negotiations between Iceland and the US were concerned, was what would happen to the meat, roughly 4,000 tonnes if they took all the scientific subjects they sought. The IWC had agreed in June that meat from scientific whales should be "primarily for local consumption"; Iceland's population of 250,000 could consume only 200 tonnes, just about 5 per cent. The rest of the whale meat would have been exported to Japan, as it had been in the past.

In the negotiations that went on during the whalers' holiday, the US persuaded the Icelanders to accept that the 5 per cent eaten there hardly constituted "primarily local consumption". Iceland, with the threat of sanctions looming, agreed to up its own consumption to 51 per cent of the scientific whale meat. A campaign to boost whale meat among the locals was instituted, but even the best marketing in the world could not get the average Icelander to eat ten times more whale meat this year than last. Instead, the remains of the whales would be fed to mink and other animals on local Icelandic fur farms. Honour would apparently be satisfied. Charles de Haes, International Director of the World Wildlife Fund, called the agreement an "immoral play with words and statistics".

But the story does not end there. By January 1987 the

Icelanders had apparently lost interest in whale meat. The Fisheries minister told parliament that their countrymen had managed to consume only 130 tonnes—less than 7 per cent of the total take in 1986 of more than 2,000 tonnes. The whale-meat campaign hurt the local sheep industry too, and as stockpiles of lamb rose, the shepherds instituted their own campaign against the whalers. Icelanders voted with their mouths, and whale meat dropped off restaurant menus, leaving the whalers with some 2,000 tonnes in cold storage.

Not all the meat, however, stayed in Iceland. Late in 1986 conservationists began to get reports from Japan that meat being shipped from Norway and Iceland was labelled "1985 catch". This deception neatly evades the agreement between the US and Japan that there should be no trade in whale products during the moratorium: "1985 catch" is before the moratorium came into effect. In March 1987 a Japanese reefer, the *Aoshima Maru*, docked in Reykjavik and took hundreds of tonnes of whale meat on board. In Hamburg, at roughly the same time, German authorities seized another consignment of about 140 tonnes which was about to be transferred from an Icelandic to a Japanese vessel. The meat was falsely labelled. Furthermore, there was far more of it than the 85 tonnes declared on the export document. Conservationists, who had been keeping a close watch on the whale meat, suspected some kind of scam, "involving official complicity, to reap excess profits and evade taxes and foreign currency controls". A second load of "hot" whale meat was en route to Hamburg but the ship was diverted to England and Belgium; when the vessel arrived in Hamburg there was no meat aboard. Conservation-ist spies are still looking for it, and these episodes indicate not only how devious the whalers can be, but also the importance of commercial trade to scientific whaling.

What is going on? Despite the moratorium, whaling continues. Of course in the worldwide context of other fisheries and, more importantly, global defence, whales are small fry; that is why whalers are permitted to ignore the decisions of the IWC. Votes are cast for reasons of political and economic expediency, and in the name of science. Private agreements allow the whalers to wheel and deal their sleazy way through any attempt to regulate the slaughter. The reason is obvious: it pays to do so.

Costs

"In one moment I've seen what has hitherto been
Enveloped in absolute mystery,
And without extra charge I will give you at large
A Lesson in Natural History."

Fit the Fifth: *The Beaver's Lesson*

The Convention that regulates the International Whaling Commission expressly states that its purpose is "to provide for the conservation of whale stocks and thus make possible the orderly development of the whaling industry". A naive view might suppose that this would not be necessary, because wise exploiters would take care not to destroy the goose that provides them with their golden eggs. Unfortunately, however, nobody owns the world's whales. That means that possession begins with appropriation; if you have captured a whale, it is yours, and if you want more whales than the next fellow all you have to do is go out and get them before he does. There is no incentive for individual whalers to restrain their depredations. Whales are a text-book example of what Garrett Hardin called "the tragedy of the commons". He explains:

> The tragedy of the commons develops in this way. Picture a pasture open to all. It is to be expected that each herdsman will try to keep as many cattle as possible on the commons. Such an arrangement may work reasonably satisfactorily for centuries because tribal wars, poaching, and disease keep the numbers of both man and beast well below the carrying capacity of the land. Finally, however, comes the day of reckoning, that is, the day when the long-desired goal of social stability becomes a reality. At this point, the inherent logic of the commons relentlessly generates tragedy.
>
> As a rational being, each herdsman seeks to maximize his gain. Explicitly or implicitly, more or less consciously, he asks, "What is the utility *to me* of adding one more animal to my herd?" This utility has one negative and one positive component.

197

1) The positive component is a function of the increment of one animal. Since the herdsman receives all the proceeds from the sale of the additional animal, the positive utility is nearly +1.

2) The negative component is a function of the additional overgrazing created by one more animal. Since, however, the effects of overgrazing are shared by all the herdsmen, the negative utility for any particular decision-making herdsman is only a fraction of −1.

Adding together the component partial utilities, the rational herdsman concludes that the only sensible course for him to pursue is to add another animal to his herd. And another; and another . . . But this is the conclusion reached by each and every rational herdsman sharing a commons. Therein is the tragedy. Each man is locked into a system that compels him to increase his herd without limit—in a world that is limited.[1]

In other words, each person making selfish use of some common resource reaps the full benefits of his selfish action, but the harm done is shared among all the other people who share the commons. Hardin explores the tragedy of the commons in many spheres: pollution, where it is far cheaper to dump wastes than destroy them; food and agriculture, where to avoid the tragedy of the commons people enclosed the common land and restricted access to wild game; and of course all these ultimately hinge on population, where the freedom to breed will bring disaster to all, even those who restrain themselves. He also explicitly applies it to whales.

> Maritime nations still respond automatically to the shibboleth of the "freedom of the seas". Professing to believe in the "inexhaustible resources of the oceans", they bring species after species of fish and whales closer to extinction.[2]

And all because whales appear to be a common resource. Whalers, like anyone who exploits a commons selfishly, face a tragedy: "Ruin is the destination to which all men rush, each pursuing his own best interest in a society that believes in the

---

[1] Hardin (1968) p. 1244.
[2] Hardin (1968) p. 1245.

freedom of the commons. Freedom in a commons brings ruin to all."[3]

To recapitulate, modern whaling began in the 1860s, when the Norwegians launched their first steam-powered whaler, equipped with a harpoon cannon. Air pumps, also developed by the Norwegians, allowed catcher boats to inflate the large rorquals and tow them back to land for processing. The explosive harpoon, another Norwegian first, made capture speedier and more efficient. As the northern stocks were depleted the whalers ventured further afield, until in 1904 the Norwegians opened a whaling station on South Georgia and began the onslaught on the Antarctic.

For the first twenty years or so Norway and the United Kingdom between them took most of the whales harvested in the southern oceans. After 1926, however, when the Norwegians first used a sea-going factory ship to process whales caught by a fleet of catchers, the rich Antarctic oceans were opened up to whalers of all nations. No longer tied to processing stations on land, the catchers were free to go where the whales were, and the take increased dramatically. In 1925 some 2,500 Blue whales were caught in the Antarctic. By 1930 the catch was more than ten times greater, nearly 30,000 whales. So vast was the haul that in 1931 the huge supply of whale oil all but drowned the market, and the whalers found it uneconomic to make the long voyage down south in 1932. Oil was consumed, and the market recovered, so that whaling could be resumed in 1933, but with a recognition now that unbridled competition could bring disaster.

Market forces ensured a more orderly exploitation of the whales. In the wake of the 1931 glut the Association of Whaling Companies drew up their own agreement to limit production for the 1932–3 season to 320,000 tons. The agreement applied to sixteen of the seventeen companies whaling in the Antarctic that season, and each company obtained a quota set in whale equivalent units. One Blue whale, which produced about 110 barrels of oil, was reckoned to be equal to two Fin whales, two and a half Humpbacks, or six Sei whales.

The catch quotas did serve to limit production and hence to

---

[3] Hardin (1968) p. 1244.

prevent wild fluctuations in the market for whale products. There were, however, two serious drawbacks to the system, especially in the years after the Second World War when shortages of edible fats made whale oil even more valuable. First, because the quotas were set in terms of Blue whale units the whalers were free to pursue whichever whales were most profitable, regardless of the state of the stocks. So a whaler who came across one of the few remaining Blue whales would, because Blue whales were the most cost-effective species, pursue and capture it. Thus the whalers increased their catch of the smaller Fin and Sei whales while continuing to inflict further damage on the already depleted Blue whales.

The second problem arose from the fact that quotas were set for the Antarctic as a whole, rather than being allocated to specific nations. The season was closed, by the Committee for International Whaling Statistics, when the agreed limit had been reached, regardless of which nations had actually caught the whales. This created intense competition to obtain the greatest share of the catch. Simply to maintain its share a company had to increase the number and efficiency of the boats it operated. The figures demonstrate this inexorable trend. In the southern summer of 1946 there were fifteen factories with 129 catcher boats; they caught 15,400 Blue whale units in a season that lasted 112 days. Five years later, in the 1951/52 season, there were nineteen factories serviced by twice as many catchers, 263. They caught 16,000 Blue whale units in half the time; the season contracted to just sixty-four days.

This state of affairs could not last, and in 1961 the whalers met, outside the auspices of the IWC, to divide the spoils. Japan got 33 per cent, Norway 32 per cent, the USSR 20 per cent, the UK 9 per cent and the Netherlands 6 per cent. The whalers could now work at their leisure, and the 1962/3 season expanded to 115 days. Private agreements between whalers are still the order of the day.

In its heyday whaling was one of the most productive oceanic fisheries, not simply in the tonnage taken but also in terms of the value of the products. At its peak—in 1931—the industry took more than 2.5 million tonnes in the Antarctic alone. Not quite fifty years later, in 1979, the total catch was less than 200,000 tons. Was this simply the inevitable outcome of the tragedy of the commons? It is traditional to assume that it was, that greedy whalers did indeed kill their golden goose, and that more enlightened international co-operation could have ensured that whaling would today still be supporting quite a sizeable industry. A detailed look at the economics, however, suggests that this alternative was never on the cards.

In a very instructive series of papers Colin Clark, a professor of Mathematics at the University of British Columbia, has shown that there is very little economic incentive for conservation. In Clark's words, "at normal interest rates whale stocks constitute a substandard investment".[4] The important word is "stocks". Clark's conclusion is based on some moderately complicated mathematics, which are fully explained in his research publications, but it is possible to gain a qualitative grasp of the theory without going too far into the equations.

With a freely competitive and completely unregulated fishery, a balance will eventually be struck between the biological productivity of the fishery and the rate of harvesting. At first, when stocks are abundant and fishermen few, the profits will be very high. These high profits attract further fishermen, who deplete the stocks and cause prices to fall by increasing the supply. At the balance, the so-called bionomic equilibrium, there will be zero profits, and no incentive for further fishermen or greater effort by those already involved. This bionomic equilibrium is economically inefficient; the fishing effort is too high and the harvest too low, relative to an economic maximum. It is stable, though, because there is no incentive to put more effort into fishing.

With appropriate figures derived from post-war whaling in Antarctica (a price of $7,000 per Blue whale unit and a cost of $5,000 per catcher per day) the balance point works out at 55,000 Blue whale units, which would provide a yield of 2,750 Blue whale units (5 per cent) each year. (This compares with a theoretically obtainable maximum sustainable yield of at least 5,000 Blue whale units.)

The yield at the theoretical balance point, a figure of 2,750 Blue whale units a year from a stock of 55,000 units, is well below the level at which whalers were actually exploiting the Antarctic at the time. That means that the whalers were

[4] Clark and Lamberson (1982) p. 104.

continuing to deplete the stocks. But it can be used to show that the model is on the right track, by comparing the effect of different prices per Blue whale unit. The figure of $7,000 is based on the Norwegian market. The Japanese, however, derived additional value from their whales because they were developing the luxury market for whale meat. In Japan, a Blue whale unit was worth about $11,000. Put that figure into the models and the bionomic equilibrium drops to 35,000 Blue whale units. Not surprisingly, then, we would expect Japan, enjoying higher prices and roughly similar costs, to continue whaling in Antarctica even after Norway and other Western nations had stopped. Such has indeed been the case.

What if Japan were the only whaling nation? Surely then it would behave prudently and conserve its stocks. "Not necessarily," says Clark, especially not if the aim is to maximise profits.

Any firm's investment decisions are based on the desire to maximise the present value of the profits it will obtain. There will be a minimum expected rate of return, usually based on the "opportunity cost" that could be obtained by putting the money instead into some equally risky alternative investment. The firm should also discount any future profit. The discount rate takes into account such things as the cost of borrowing money and likely inflation, as well as the risks of the business, and it becomes particularly important in investment decisions to exploit natural resources.

Imagine that you own some renewable resource, say a school of whales, and that you wish to exploit it for maximum profit. You are faced with a choice between milking and mining: you can conserve your asset forever while harvesting only the sustainable yield, or you can sell off the whole resource for an early profit which you then invest in some other venture. (Intermediate solutions are also possible, but we will ignore them.) If you conserve the stock and harvest the yield, your income might be £V per annum. If instead you cash in the whole stock for a lump sum of £P and invest it elsewhere with an annual rate of return of i per cent, then your annual income will be £P × i. Your strategy then depends on the relative values of P, V, and i. If P × i is greater than V, your best option is to cash in the stock and invest elsewhere.

Vs, Ps and i are all very well, but the idea can be brought to life with figures from Antarctic whaling in 1946. Suppose there were 200,000 Blue whale units swimming about, providing an annual sustainable yield of 5 per cent, or 10,000 Blue whale units. With each Blue whale unit worth $7,000, a sustained harvest yields $70 million a year. If, however, you could capture the whole stock and liquidate it at the same price of $7,000 per Blue whale unit, you would have a lump sum of $1.4 billion. Invested at a rate of 10 per cent, your return would be $140 million a year, twice as much as you might get from sustained harvesting.

You may think that this model is too simplistic, but even when the costs of building the fleets and suchlike are taken into account the basic conclusion remains. The crucial point is the relationship between the rate of return on whales as opposed to the rate of return on money. If whales multiplied more quickly than money, milking would be a better bet than mining. The fact of the matter is that the figures I have quoted are probably too conservative. Whale stocks almost certainly reproduce much more slowly than 5 per cent per annum; 1 or 2 per cent is probably closer to the norm. And monetary investments often provide considerably in excess of 10 per cent. So long as that is true mining will make more money than milking; whale conservation will never pay.

This is a pretty pessimistic conclusion for anyone with a non-financial interest in whales. It means, in effect, that we cannot expect whaling nations to regulate their activities "rationally", because in fact the rational approach is to exploit the stock as quickly as possible until it is exhausted. That may be disappointing, but it is true. All the so-called disasters of the history of whaling—the system of Blue whale units, the depletion of the most valuable resources first, the excessive fishing capacity of the early heady days—are in fact entirely rational responses to the motive of maximising profits.

Against such a background the idea of saving the whales seems positively absurd. The money to be made in whaling guarantees overexploitation. For that reason alone, a sustainable harvest is not an option to investors in whaling. Without knowing this, however, a small group of people set out to tip the balance sheets in favour of the whales.

Until just fifteen years ago, almost the only people with any interest in whales were whalers. For the vast majority of the public these enormous creatures remained shadowy and enigmatic. They were certainly not the kind of animals with which one could easily empathise. Indeed very few people ever saw a whale. They did not realise whales were mammals. They did not realise the extent to which whales were being killed. Today, in 1987, whales are big box-office. "Save the Whales" is now a global rallying cry, as familiar in remote villages in India as it is in conservation-crazy California. Many people now seem convinced that whales are immensely intelligent beings, probably more so than humans. Those people turned a movie predicated on that fact—*Star Trek IV*—into an enormous success worldwide. A television documentary about whales is the most successful programme the National Geographical Society has ever made. That whales are such bankable stars is testimony to the remarkable campaign waged on their behalf.

There had been conservationists before, of course, but mostly they were caught up in ways of managing nature, manipulating it for the benefit of some species (mostly *Homo sapiens*) and arrogantly imagining that they could understand enough of nature to manipulate it successfully. The new-style conservationists, mostly young Americans fresh from the campaign against the war in Vietnam, understood that they could manipulate people and governments more easily than ecosystems. Whatever their various motivations, they set about protecting the environment in all sorts of ways, and one of their greatest successes was the US Marine Mammal Protection Act, passed in 1972.

With all the whales being killed for commercial reasons, it is astonishing to realise that at one stage more than a million Cetaceans—admittedly somewhat smaller than the great whales—were being wasted each year. They were not killed by accident, they were deliberately drowned, but no use was made of them at all. They were the dolphins that fishermen took advantage of to find the huge shoals of money-spinning tuna.

The tuna industry had been created in the 1940s, almost entirely by Yugoslav and Portuguese families in southern California. At first the fishermen used lines attached to poles, several poles on every boat. Although these tools seem quaint to today's fisherman, they caught very many tuna. At the end of the Second World War canned tuna, so ubiquitous today, was almost unknown, but clever marketing quickly created a demand. The canneries of the California coast expanded and grew to process the fish, which soon became a very profitable industry. In the late 1960s the approach changed, and some of the tuna boats started using purse seine nets to gather up the shoals. It was then that the skippers began to exploit the association between dolphins and tuna.

Fishermen had always known in the past that tuna often kept company with dolphins. Now they realised that the tuna actually followed the dolphins, perhaps because the Cetaceans were so much better at finding the fish that tuna and dolphins both feed upon. The fishermen began to set their nets deliberately around schools of dolphins. They caught huge numbers of tuna. They also trapped the dolphins who had shown them where the tuna were. The dolphins drowned.

Craig Van Note, head of Monitor Coalition, a consortium of conservation bodies in Washington DC, insists that the tuna fishermen "destroyed the resource, and the dolphins too. They killed over a million dolphins in one year. These guys were a bunch of pirates." The tuna skippers had always been a secretive lot. Now onlookers would see a boat head out to sea only to return just a couple of weeks later, her holds bulging with tuna. It did not take long before everyone was in on the act, building new boats at a cost of $10 to $15 million. For a while the tuna business was very valuable; the boat owners made their investment back in a year or two, skippers earned $500,000 a year, and even ordinary fishermen were pocketing $100,000.

Of course these profits could not last; such overexploitation never does. But during the boom years it was dolphins that lost out. Nobody will ever know how many dolphins drowned in tuna nets. Van Note publicly accused the industry of having killed 6 million animals. "They never challenged that figure," he says, "though myself I think it could have been closer to 10 million."

The public knew nothing of the dolphin slaughter until very late in the decade, and the figures make grim reading. In 1960, the worst year, an estimated 853,000 dolphins drowned in tuna nets, about 350 each time the skippers set their nets for tuna. The side effect of their appetite for canned tuna was brought home to people with bumper stickers that read "Would you kill Flipper for a tuna sandwich?" By 1972 opinions had shifted enough to allow conservationists successfully to introduce the Marine Mammal Protection Act. But the government department responsible, then called the Bureau of Commercial Fisheries, now the National Marine Fisheries Service, refused to have anything to do with the new legislation. The Bureau allowed tuna fishermen an unlimited catch of porpoises for 1974 and 1975. Van Note's Monitor Coalition sued, but it was not until 1977 that the final decision emerged, when the Supreme Court refused to hear the government's appeal. A lower judge had decided that the "maintenance of a healthy tuna industry was not one of the [National Marine Fisheries Service's] responsibilities under the act".[5] A reluctant NMFS forced regulation on the tuna fishermen.

At that stage about 100,000 dolphins were drowning each year; the new legislation introduced a special net design, including panels that would allow the dolphins to escape. It also required fishermen to get themselves trained in the art of saving dolphins, and devised a manoeuvre called backing down, in which the boat reversed towards the end of the net, submerging the net and giving the dolphins the chance to get out. Publicity films at the time show friendly fishermen helping their dolphin assistants out of the net. The reality was somewhat different.[6]

The fishermen complied when they were being watched, which was only about 30 per cent of the time. One government observer called a halt to the fishing when the skipper asked him to falsify his tally sheets so that they would show fewer dolphin deaths than had actually occurred. Others reported that the men who were supposed to help the dolphins out of the net did little or nothing. The government observers' reports indicate that the tuna boats hardly complied with the letter, never mind the spirit, of the new law. Perhaps worst of all, for a while the skippers indulged in a practice called "sundown sets". They paid out the nets as dusk fell, and as they gathered in the harvest later that night they could legitimately claim that they had "not knowingly killed any marine mammals"; it had been

[5]Federal Register, 42 (247), 23 December, 1979.
[6]Most of these examples are from *Whales vs Whalers: A Continuing Controversy.*

too dark to see what exactly they had killed. Sundown sets are now illegal, unless the vessel has high-powered arc lights to illuminate the scene.

Before they accepted these measures, however grudgingly, the tuna fishermen went on strike, tying their fleet of seventy-five ships to the docks in San Diego and San Pedro. They flew planeloads of fish wives into Washington to lobby Congress to gut the Marine Mammal Protection Act, but the women's campaign made a very bad impression. Perhaps that was because their lamentations of impending economic disaster could hardly be heard over the jangling of their massed gold jewellery. For whatever reason, the strike gained nothing, and probably lost the industry tens of millions of dollars. In any case, by 1977 the catches were beginning to fall and the American tuna industry was reduced to a fraction of its size just fifteen years earlier.

What is the future for tuna, dolphin, and fishermen? The skippers say they have mended their ways, but more than 90 per cent of the yellowfin tuna caught by the US fleet were still taken by setting nets around schools of dolphins. The American Tunaboat Association has instituted a Golden Porpoise Award for the skipper with the fewest dolphin kills in a season, though whether this will tempt skippers to forgo a few extra tonnes of tuna remains to be seen. The kill rate is down, from its peak of 350, to just eight dolphins for every haul of tuna. Nevertheless, in 1986, for the first time, the tuna boats reached their kill quota of 20,500 dolphins before the end of the season. The National Marine Fisheries Service ordered the fleet not to catch any yellowfin or bigeye tuna unless the boat had an observer on board to see that nets were never set on dolphins. This order was rescinded on 1 January, 1987, when the tally started afresh.[7]

The eastern spinner, the dolphin the skippers once set their nets around, is now officially listed by the US government as a depleted species. That means the tuna boats are supposed to ignore schools of eastern spinners. As a result they have shifted their attention to offshore spotted dolphins. Often spinners and

---

[7] In 1987, for the first time, every tuna boat will have a government observer on board. One skipper sued the government when he discovered that his assigned observer was to be a woman. He said the presence of women disrupts life on tuna boats designed for all male crews. The courts have not yet decided the issue.

spotted dolphins form mixed schools, beneath which the tuna still lurk. Such a mixed school of Cetaceans might make the tuna skipper's job harder if he really wanted to comply with the law. Some conservationists are now talking about a campaign to get the government to recognise that stocks of offshore spotted dolphins too are now depleted. If they succeed the American tuna industry would probably collapse. The fishermen say they would get the Marine Mammal Protection Act ditched or changed, but Van Note and others think it unlikely that they would succeed. But with the home industry so diminished, American conservationists were worried that other nations would step in to fill the gap in the canned tuna market, and those fishermen won't be hampered by any need to save dolphins while slaughtering tuna. As a result the US now has an amendment to the Marine Mammal Protection Act which requires nations that export tuna to the US to have similar standards for dolphin protection—but it is not being enforced.

The dolphins too have not been idle. "They've learned that tuna boats mean death," says Van Note. They take off like a shot when they first detect the special sounds of a fishing boat, and the boat may have to chase them for miles. The school often splits into two or three portions, forcing the skipper to commit himself to one, but when he finally catches up with the dolphins, for eventually they must tire, and sets his nets around them, he still comes up with a good haul of tuna. The fish are clearly not as smart as the Cetaceans; for whatever reason, the tuna follow the dolphins, and stay with them.

The behaviour of the tuna might itself provide a solution, because dolphin schools are not the only objects that have tuna beneath them. Almost any large object floating in the eastern Pacific will have tuna under it, and boats often set their nets successfully around a palm tree bobbing on the waves. Perhaps it will be possible in the future to develop large floating rafts that might be taken to likely spots and left there to gather a good shoal of tuna. In the meantime, dolphins still die in American tuna nets, at a rate of 20,500 a year. Mexico is killing more than 100,000 a year. How many the other Pacific nations—Korea, the Philippines, Japan—destroy remains an enigma.[8]

---

[8] A new threat is posed by the almost invisible drift nets, curtains of transparent monofilament, miles long, that fishermen pay out into the sea. These nets catch animals indiscriminately, and conservationists are working hard to get their use restricted.

The Marine Mammal Protection Act was the first great success for the new conservationists, and it was just one expression of changing attitudes, especially in California. *Silent Spring* had alarmed people, ecology was the new buzzword, expanded consciousness encompassed whale brains and the rights of trees. The first battle against the tuna fishermen had been won. Conservation mattered, and whales were the biggest of the beneficiaries. Educators along the coast were using whales as a monstrous peg on which to hang all sorts of ideas, from the inter-relatedness of the species in an ecosystem—whales eat krill and men kill whales—to the influence of political systems on their subjects—who should control the fate of international animals like whales. They were taking children out in small boats to see the Gray whales on their migrations, and before too long the children had fired up their parents.

This, crudely, was the background to the United States' participation at the 1972 United Nations conference in Stockholm. Delegates were supposed to consider the current status and future of the Human Environment. Fresh from the passing of the Marine Mammal Protection Act, the US sought to extend that protection to all whales, everywhere. The Stockholm conference was asked to approve a ten-year moratorium on whaling. A pause, not a cessation. The delegates agreed overwhelmingly.

Two weeks later the United States delegation to the IWC arrived at the meetings in London and presented the same proposal. As Van Note remembers it, "of the 13 member nations 8 or 9 were whalers. They literally laughed in the face of the US and told us to take a walk. The commissioner was so furious he vowed that the US would take on this fight."
While government did its best, the handful of conservationists who were worried about whales met in the autumn of 1973 in Washington and decided to launch the great Save the Whale campaign with a boycott against Japan. The boycotts gave the public a way to vent their anger, but otherwise did not do much good. The rest of the campaign surely worked.

A major strategy was to purchase advertising space all around the world, seeking to spread the message of whales, dispelling some of the enigmas and heightening people's awareness. One organisation, the Animal Welfare Institute, "probably spent $1.5 million purchasing whale ads, most of which I wrote," Van Note claims. "I think I've written more than 200 Save the Whale ads which we've run in every major English language publication in the world and quite a few foreign language publications." A slight exaggeration, but typical and perhaps justified. "I've been called the Goebbels of the conservation movement," Van Note says with some pride, "and I take that as a compliment." The campaign has made people around the world, 99.9 per cent of whom will never see a whale, think of whales as part of a global heritage.

Perhaps the most memorable of all the Save the Whale efforts were the intrepid exploits of gung-ho Greenpeacers. They might not be able to stop whaling governments, but they could certainly interfere with whaling operations. They decided, paraphrasing Lytton Strachey, to interpose their bodies between the whalers and the whales. Tiny inflatable motorboats —Zodiacs—buzzed about like demented flies beneath the massive bows of the catcher ships. The protestors tried to keep themselves between the harpooner and his quarry, so that any shot fired at the whale would risk injuring the humans below. By and large the whalers did show more respect for human lives than for the whales', although there were a number of near misses. In the first such confrontation, off the California coast in 1975, a Soviet harpooner fired at a Sperm whale cow over the heads of one Zodiac.

The next year, Soviet harpooners abandoned their weapons whenever the protestors took to their Zodiacs, but there simply were not enough protestors to protect all whales everywhere. Nevertheless, the Soviet fleet abandoned its practice of whaling in the eastern Pacific, and never came closer to the west coast of North America than 700 miles. Becoming more militant still, in 1977 Greenpeacers drove their inflatables up the slipway of the *Dalniy Vostok*, a factory ship, and harangued the Soviet crew. They took on the Australian whaling station, in what proved to be its last year. People in Zodiacs seemed to be everywhere.

Much of Greenpeace's apparent success can be ascribed directly to their fine grasp of the power of pictures; they know a good story when they see one. Greenpeace always made sure that there were professional photographers on board, and news bulletins showing little Davids taking on the mighty whaling Goliaths brought the campaign into millions of homes. In 1983 the *Rainbow Warrior* sailed to Siberia to investigate the industry there. The IWC lets the Soviet Union catch 179 protected Gray whales there for the aboriginal Eskimos to eat. Sea Shepherd, a breakaway group (see p 213), had two years earlier claimed to have evidence that the Soviets were feeding their special allowance of Gray whales to mink on commercial fur farms. Nobody, however, had heard of that effort.

When Greenpeace landed at the Soviet whaling station near Loren, the emphasis was on pictures. After a dramatic chase across the high seas, some of the protestors were captured and equipment smashed. Satisfied with this heavy-handed punishment, the authorities returned the protestors to their ship. The Soviets, however, had not been very thorough in their search of the Greenpeacers. One woman had hidden a roll of film in her vagina. The resultant photographs—along with Greenpeace's film of the harrowing chase across the Arctic seas—made this "one of Greenpeace's best media-covered campaigns ever".[9]

While Greenpeace was busy organising media events, petitions, boycotts and the like, other protestors were saving whales in other ways. Friends of the Earth mobilised public opinion in Great Britain, but also campaigned behind the scenes for an important shift in policy. Britain had banned the import of most whale products in 1973, in the wake of the Stockholm conference. The exception was Sperm whale oil. This liquid fat was unsurpassed as a lubricant in machinery —US atomic submarines and Soviet tanks equally depended on it—and it was also used in many other products. Britain represented a major clearing house for Sperm oil, with an important refinery in Scotland, and Friends of the Earth decided to work hard for a complete ban on Sperm oil.[10]

Lists of those companies using Sperm oil were circulated, and a few consumers may have boycotted them. Much more important, campaigners from Friends of the Earth undertook to educate the users, a policy that quite quickly paid off because "many firms were genuinely unaware that the whale was endangered and that they were contributing to the whales' demise".[11] Substitutes were found that could stand in for Sperm oil in every one of its many uses. Friends of the Earth

---

[9] Greenpeace anti-whaling campaigns, 12 February, 1986, p 8.

[10] In 1978, Britain imported some 4,000 tonnes of sperm oil, worth about £1½ million. Almost all of this was re-exported, mostly to the rest of Europe.

[11] Anon. (1978) p 115.

turned its attention to "intensive lobbying of a few key users". These included the Ministry of Defence and Clarks Shoes, both of which played important parts in the next step, getting the government to agree a complete ban on the import of Sperm oil, both in bulk and as a component of other products.

One of the biggest users of Sperm oil in Britain at the time was the Ministry of Defence, and much of Friends of the Earth's "intensive lobbying" was directed at the minister, John Gilbert. The Ministry received some 3,000 protest letters, and some very carefully researched arguments. Gilbert was eventually convinced that there was no need for Sperm oil in the armed forces, and agreed to address a Save the Whale rally in Trafalgar Square in the centre of London, the day before the 1979 meeting of the IWC[12]. The day was grey, but Gilbert's speech was the brightest moment in a long campaign. Explaining his decision to use no more whale products, Gilbert said to the crowd, "I can promise you that no more RAF fighter pilots will ever fly wearing gloves treated with Sperm whale oil." As Gilbert flung his arm skyward for emphasis, three or four fighters zoomed overhead. It was an astonishing piece of theatre. John Halkes, chairman of Friends of the Earth at the time, had been an RAF pilot himself. When Gilbert sat down, to rapturous applause, Halkes leaned over and ventured that it must have taken months of training to get the display so precisely perfect. Pure coincidence, said Gilbert, who knew perfectly well that no amount of planning or training would ever have allowed the RAF to demonstrate at a Save the Whale rally.

There was, it seemed, no stopping the campaign, but conservationists were disappointed when the British government failed to announce a ban next day at the opening session of the IWC. The Ministry of Defence, government's biggest consumer, might have abandoned Sperm oil, but there were bigger fish to fry. Tom King, minister at the Department of the Environment, saw this as a wonderful issue for Britain to take to its partners in the European Community. Things were not entirely easy on the home front either. UK Customs and

---

[12] The invitation to Gilbert caused a little rift in the Save the Whale campaign; Greenpeace (London) "formally dissociated" itself from the rally, unable to share a platform with John Gilbert, whom it quaintly described as a "Minister of War" from the "Ministry of 'Defence'". (Press release, 7 July, 1979.) According to Tom Burke, "nobody noticed, not even Greenpeace Vancouver".

Excise, which would have been responsible for policing any such ban, at first demurred. Inspectors said they would be unable to detect Sperm oil when it had been used, for example, to treat leather. Clarks Shoes, however, had decided not to use any Sperm oil itself, nor to buy any leather treated with Sperm oil. To implement this decision it needed a way of detecting Sperm oil in leather from outside suppliers, and scientists at Clarks quickly developed an accurate and sensitive test. Clarks offered the test to Customs and Excise; whether the offer was ever taken up I do not know, but one objection to the ban fell. Finally, after a year of hard bargaining, the Europeans agreed to ban all whale products, a significant step forward that was announced just before the 1980 meeting of the IWC.

After the European ban on Sperm oil and all other whale products, it seemed just a matter of time before one of the moratorium proposals passed. Effective campaigning by groups such as Monitor ensured that the American government kept pressing for a ban, bringing erstwhile pirate whalers into the fold and keeping the IWC afloat. Greenpeace continued to make waves, although all the important decisions had already been taken. The whaling, however, continued, and some conservationists had again grown weary of what they saw as a softness in their fellow campaigners. They preferred to take direct action. Thus it was that on 8 November, 1986, Rodney Coronado and David Howard, two volunteers from the Sea Shepherd Conservation Society, slipped ashore in Iceland, intent on destruction. Their first stop was the whale processing station at Hvalforjdur, where they broke in and sledge-hammered all the computers and other equipment they could find. They also switched off the refrigeration on a warehouse that contained 2,000 tonnes of whale meat; the saboteurs said this meat was packaged in cardboard cartons clearly destined for Japan.

Leaving the meat to thaw, the two drove the fifteen miles south to Reykjavik, arriving in the early hours of Sunday morning. In the harbour, three of Iceland's four whaling ships were moored alongside one another. The saboteurs made their way onto the outer two vessels, checked that there was nobody on board, and loosened the bolts around the seacocks. Water flooded into the engine rooms, sinking the *Hvalur 6* and *Hvalur 7*.

One night's dangerous work had cost the Icelandic whaling industry half of its fleet, and something in the region of $2.5 million in repairs. After a narrow escape from a routine police block, the pair boarded a flight to Luxembourg.

The raid on Iceland was just the latest piece of "ecotage" by the Sea Shepherds. The movement had begun in 1977, when its founder, Paul Watson, was expelled from Greenpeace. Watson had been a founding member of Greenpeace, but his methods—grabbing the weapon from a sealer busy clubbing pups and hurling it into the sea—were not peaceful enough for Greenpeace. Watson quit in an acrimonious parting of the ways, and launched the *Sea Shepherd*, dedicated to direct action in the conservation of marine wildlife.

It was the *Sea Shepherd* that achieved what international persuasion had failed to, bringing the career of the pirate whaler *Sierra* to an end. On 16 July, 1979, Watson deliberately rammed the pirate twice. His own ship had been reinforced with concrete for the attack, and suffered little damage. The *Sierra* limped back to Lisbon for six months of costly repairs. On 6 February, 1980, the Sea Shepherds struck again, sinking the *Sierra* in the harbour. Later that year, in April, they sank another two ships, the *Isba I* and *Isba II*, Spanish whalers in the northern port of Vigo. And in August 1981, the *Sea Shepherd* violated Soviet law to obtain evidence that the USSR was using its take of 179 Gray whales, permitted by the IWC for aboriginal subsistence, as a source of cheap protein for mink farms along the Chuckchi coast.

Watson admits that he breaks the law, but only, he says, to uphold what he sees as a greater law. If the IWC cannot enforce its decisions, Sea Shepherd will try to do so for them. "Although we are not an authorised policing body," Watson is fond of saying, "we are a policing body." Greenpeace is still feuding with Watson. Alan Reichman, a spokesman for the organisation, says that "violence doesn't work". Greenpeace and Sea Shepherd are ostensibly aiming at the same targets, but Reichman worries that Greenpeace's efforts will be tainted by Sea Shepherd's brand of vigilante conservation. He also sees little difference between Sea Shepherd sinking a pirate whaler and the French secret service sinking Greenpeace's own vessel *Rainbow Warrior*. "Those things happen when you believe you can act outside the bounds of international law," Reichman explained.

Other conservationists tend to equivocate about such direct actions. Sidney Holt, speaking at a meeting of the American Cetacean Society just two weeks after the Icelandic scuppering, said: "I hate in any way to appear to support terrorist acts, because I don't . . . yet the end result is there are now two less whaling ships in the world." That, in the end, is Watson's justification. While others talk and make deals, he stops whalers. By very directly upping the ante, Sea Shepherd forces the whalers to reconsider their profit and loss accounts.

Greenpeace's spectacular derring-do certainly alerted people to the Save the Whale campaign. Whether it saved the whales is a much more contentious point. I believe it did not, for two reasons. One is that Greenpeace itself joined the fray when the political battle had been all but won. As Sidney Holt put it, Greenpeace "came rather late to the scene of trying to stop the depletion of whales—years after the Stockholm UN conference, for example". Because of their firm grasp of public propaganda, however, Greenpeace succeeded not only in putting it about that the whales had been saved, but also that they, Greenpeace, had saved them. The brave volunteers in their Zodiacs did save the lives of a few individual whales, and I would not want to diminish their achievement. But Watson's Sea Shepherds saved more whales directly, and credit for the perceived success of the Save the Whale campaign is too often ascribed in error to Greenpeace. Tom Burke, at the time the leading light of the Friends of the Earth campaign, is quite blunt about it. "Greenpeace have tried to rewrite history," he says, "presenting themselves as the people who saved the whale, but the claim just doesn't hold up."

Such revisionism would hardly matter if the whales had in fact been saved. The fact is that they have not been, not by any means. The die-hard whalers may have changed the name of the game from economic harvest to scientific research, but the truth is that they are continuing to whale more or less as before. Nevertheless, it is curious how many people who, hardly knowing what a whale was just fifteen years ago, now know not only what they are but also think they have been saved.

There is another point here. It is that the conservationists, in

all their guises, may have prolonged the agony of the whales
hunted by industry. David Day, in his version of the Save the
Whale story, says "it was quite apparent that under its own
steam the whaling industry would have died out in the late
1960s."[13] The history of whaling is an inevitable series of booms
and busts. Each time a new population was discovered, it was
overexploited. Only increasing technology enabled the whalers
to survive. Having dug deep into the last remaining populations of whales, the Antarctic stocks, it is hard to imagine
what new bits of technology could have saved the industry if it
had collapsed. But it was not allowed to collapse.

Extinction is a funny thing. In a sense, it is very sudden. The
last passenger pigeon, for example, died one day in 1914, and
that was the end of that species. But before that happens, there
is a lingering decline. The California condor has been slipping
slowly from its stronghold in the mountains around Los
Angeles for the past century or more, although the last wild
individual was not captured and brought into captivity until
the spring of 1987. The plan is to breed the birds in captivity,
one day to release them back into the wild. Whether that plan
succeeds is not the point. It is that long before a species becomes
extinct, it ceases to be exploitable. Economic extinction, unlike
its biological counterpart, happens quite suddenly.

Imagine the giant whaling fleet of the 1950s, scouring the
Antarctic for prey. In 1964 a hundred Japanese catcher boats
could not find a single Blue whale, and it was after this
staggering failure that the IWC finally gave the Blue whale full
protection. What if there had been no protection, no quotas, no
agreements between the whaling nations? Is it not possible that
all the fleets would one year have failed to find enough whales to
cover their costs? And when that happened, as I believe it
surely would have, the investors would quickly have pulled out
of whaling.

Apart from their customers, the people who supported commercial whaling were not the men on the boats. Support came
from the men with the money to buy the boats, which they did
because it was very profitable. As soon as it became unprofitable, they would have abandoned whaling in a hurry. Instead,
protest and protection within the IWC gave the money men

[13] Day (1987) p 116.

time to shrink their fleets, to reduce their investment slowly so that it remained profitable longer. During that time, because a whaling industry still existed, there was also a ready market for the whale products obtained by pirates. One of the conservationists' greatest successes was in uncovering the connections between Japan and the pirate whalers, as a result of which Japan ostensibly stopped importing whale products from non-IWC countries.

Now shrunk almost, but not quite, out of existence, the whaling operations survive on subsidies. The USSR announced in 1985 that it would cease whaling in the Antarctic after the 1987 season. The fleet is rusted through and it is not worth replacing. Japanese whalers are an enormous drain on their parent companies, maintained now for political reasons rather than sound economic sense. Norway's intransigence owes more to the powerful fishing lobby, and a desire to avoid unemployment, than any need to whale. Iceland, Korea, the Philippines, all are squeezing the last drops of profit from their investment in whaling. Commercial whaling will stop, not when the protestors demand it, nor when the IWC decrees it, but when it no longer pays.

There were, as I see it, two alternatives in the 1960s, when it became apparent that the whalers were overexploiting their final stocks. Conservationists could have done nothing: I believe the industry would then have collapsed under its own economic weight very quickly. Or they could campaign to save the remaining whales, as indeed they did. I believe it possible that more whales died as a result of that decision.

Part of the problem was that at the time it seemed foolish to advocate that whalers completely abandon their prey. Despite the very familiar excesses of the past, it seemed far more sensible to aim at rational management that skimmed off the surplus, leaving the stocks intact. That was what the whalers said they wanted, the maximum sustainable yield, and the conservationists agreed that this was a reasonable goal. Many people still think that it could be possible to exploit whales forever, if only the whalers were not quite so greedy.

Sidney Holt, one of the original committee of three scientists who advised the IWC on quotas after the crash of the 1960s, has consistently pointed out flaws in the management approach to whales. He now even doubts whether there is such a thing as maximum sustainable yield. And yet he succumbs to the lure of endless exploitation. Arguing the need to protect the Minke whales in the Antarctic immediately, even though the catch there is still relatively healthy, he wrote me a letter saying that, "for that one stock of one species . . . there is just a small chance that the nations could get their act together in time to have a small but continuous whaling industry." He went on to admit that "this is unlikely to happen because economics is against it, and the IWC is politically miniscule (sic)".

Economics is against it, because money multiplies faster than whales. And the IWC is bereft of power. But Colin Clark's analysis of the economics of whaling does suggest a way out. The problems stem from the fact that nobody owns the world's whales. Colin Clark proposes that ownership be vested in something he calls a World Whaling Authority. It could just as easily be a restyled IWC, but given the birth of the IWC as a gathering of whalers, perhaps something more neutral would be better. This World Whaling Authority might be an offshoot of the United Nations, and at the stroke of midnight on, say, 1 January, 1990, all the world's whales would belong to it.

The WWA would be required by its charter to conserve whales, but could sell annual quotas to the highest bidders. The income would go to running the Authority, and so the price might be very high, but this would be a realistic value for a whale in the water. Among the bidders would be whalers, putting a value on their product at the margin, where it counts. They know how much profit there is to be made in processing a free whale, and would have to decide how much the raw materials are worth. But conservationists would also be free to bid, replacing rhetoric and campaigns with responsibility and cash. The conservationists could do as they pleased with their quota; they might choose to harvest some for additional funds, or they might exploit their quota benignly by encouraging whale-watchers. Or they might do nothing, simply allowing the whales to multiply.

Clark admits that his proposal is fanciful, but that does not mean it is foolish. Conservationists might find themselves outbid at every auction. Or perhaps putting a price tag on each whale is all that's needed to halt commercial whaling forever. The World Whaling Authority would need a lot of clout to enforce its ownership, clout that would presumably have to be financed at least in part by the fee income it obtained for quotas. Nobody knows whether it would work. Nevertheless, as a possible happy ending to the tragedy of the commons, it certainly bears consideration.

Epilogue

"My poor client's fate now depends on your votes."
Here the speaker sat down in his place,
And directed the Judge to refer to his notes
And briefly to sum up the case.

Fit the Sixth: *The Barrister's Dream*

Whaling as a harvest cannot pay, certainly not at the levels the whalers have come to expect. With the remaining whales in the ocean available for free, it will always pay people to pursue them for immediate short-term gain. The history of whaling shows that clearly. In every single instance, commercial whaling has led to the decline, and ultimately the collapse, of the stock. From the Basque whalers of the fifteenth century to the Antarctic of the twentieth it is the same story. Deplete one stock to the point of economic extinction and move on to the next. The breaches of the moratorium are simply an expression of that fact, an attempt to wring the last drop from the resource.

Technology has come to the rescue over and over again, enabling whalers to exploit new stocks that were still pristine only because nobody had been able to use them. Now, I suspect, we are near the end of the line. The Minke is the smallest great whale, and although there are small industries based on other Cetaceans—Pilot whales, killer whales, beaked whales and the like—they do not have the global destructive power of the modern whaling fleets of twenty years ago.

How did we get into this position? There are several points at which, were we able, we might choose to rewrite history.

Perhaps the first is that nobody realised, until too late, just how slowly whales reproduce. As a result of their initial overwhelming abundance, whales provided a tempting target. Because they were common to all mankind, no group of men could be expected to withdraw from the bounty. In the scramble to exploit the whales, we took too many, and by a considerable margin. Had we known, back in 1920, even the

221

little we know now, we might have been able to set up a sustainable whaling industry.

When, finally, the whalers themselves saw the need to regulate their activities there was an exceedingly naive assumption that they would be gentlemen who would abide by the rules. The International Convention for the Regulation of Whaling, 1946, is, like the baleen whales, entirely toothless. There are, it is true, financial penalties for whalers who break the rules, but these are trivial when imposed, and more often than not no penalties are imposed, even when the country concerned admits its infractions. Apart from that limited power, the IWC has no force. It cannot require any nation to belong. It cannot prevent any nation ignoring its decisions. How is it to regulate for the long term when there is nothing to lose, and everything to gain, by ignoring the IWC in favour of short-term profits?

The only sanctions in existence are outside the IWC and outside whaling. Political opprobrium may hold some sway over some miscreants, but it has a way of melting in the heat of other, non-whaling considerations. As Norway and Iceland prepared to ignore the IWC moratorium one could hear dark mutterings about the continued existence of NATO, obviously more important than the continued existence of a few whales. Public feeling is important, but the kind of government that is prepared to carry on whaling today is not obviously open to persuasion by public feeling. Norway's hypocrisy is especially hard to condone.

Before she became prime minister of Norway for the second time, Gro Harlem Brundtland was appointed chairman of the World Commission on Environment and Development. The United Nations had set up this august body with an urgent call to formulate "a global agenda for change". The report of the commission, launched in London with all appropriate fanfare and ceremony in April 1987, is a fat document full of well-meaning resolutions. It acknowledges that "many species and fishing grounds have been overexploited. The seas are in trouble." It is called Our Common Future, surely more than a gesture to the obvious fact that so many of the perils that threaten us are indeed the seemingly inescapable tragedies of the commons. Mrs Brundtland announced that the purpose of Our Common Future was "to make the world aware that humanity has come to a crossroad", and yet her own tragedy rolls inescapably on.

The Norwegians have had many perfect opportunities to stop whaling. They could have done so when the scientists insisted the stock be given protected status, in 1985. Mrs Brundtland could have done so in 1986, when she came back into power. She could even have done so as a political gesture, to accord with the theme of Our Common Future. Instead, the week before the launch of the report, her fisheries minister announced the quota for the 1987 kill of Minke whales in the North Atlantic. While announcing that the rest of the world faced a succession of crises, Mrs Brundtland was unable to prevent her own contribution to the tragedy of the commons, not a very hopeful sign. Of course she was challenged on this point at the press conference after the launch, but she merely repeated the assurance that Norway would stop commercial whaling in 1987. She made no mention of the fact that Norway intends to continue with scientific whaling in 1988.

There are two reasons why Norway and the others continue whaling today. The first is that to stop would deprive a few people of part of their income. It should not be beyond the wit of a democratic government to find some way of getting around this problem. Nobody absolutely depends on whaling for their livelihood. The other, much more important, is the money that exports of meat to Japan bring. If the Japanese would abandon whale meat, the entire global commercial whale hunt would stop tomorrow. But the Japanese seem unable to do this. They smile inscrutably, and say "Yes" to avoid insulting you, but all their behaviour says "No". The Japanese whaling industry refuses to give up, and at present plans call for a scientific haul of 825 Minkes and fifty Sperm whales every year for at least twelve years. The Ministry of Agriculture, Fisheries and Food is supporting this research with a subsidy of Y350 million. The Japanese obviously hope that the industry will be able to limp along until the moratorium is reassessed in 1990, when they will attempt to use their scientific data to re-open commercial whaling. Ever pragmatic, however, one Tokyo restaurant that specialises in whale meat has set aside storage space for 70 tonnes of frozen Minke meat. That is equivalent to about fourteen whales, and is enough to keep the restaurant in business for at least the next six years.

The whalers seem unwilling to respond to mere opprobrium. Economic efforts may be more fruitful, and consumer boycotts can achieve limited goals. Government boycotts might be more effective, but they are far harder to impose. Certainly the only government with the power to affect whaling through economic sanctions, the United States, has chosen not to do so.

Lack of effective weapons was not the only problem with the IWC. The Blue whale unit, first mooted in the 1930s, probably hastened the demise of the stocks. It failed to protect the whales that needed protection most, because whalers were free to go after the most profitable species, which by their very nature were those that were most severely overexploited. Without complete protection, those species could never hope to recover. The Blue whale unit might have been more successful if it had been used more sensitively; if, for example, it had been adjusted to take account of market forces and the biological condition of the stocks. Making a Blue whale equal to two, or even three, Blue whale units might have decreased the value of the animal to the point where a catcher would pass by a Blue whale in search of Fins or Seis. Unfortunately, not only was the Blue whale unit fixed for all time in 1931, thus taking no account of subsequent changes in the value or status of each species, it also lingered far too long. If it had been scrapped earlier, that too might have prevented further decline.

Another tantalising "might have been" concerns the quotas themselves. What if the IWC had started reducing quotas sooner and more gently? The actual quota for the Antarctic was finally chopped by a third in 1963. Recommendations for a smaller take had been on the table several times before, but had never been accepted. The reduction, when it came, was too late and too savage. If the whalers had seen the light earlier, and agreed that, while stocks might not be able to support the then current whaling effort, they could perhaps support a reduced fleet, they might have wound down their activities slowly and sensibly to an acceptable level. Instead they were faced with an industry too big to survive on the allotted number of whales. Rather than biting the bullet and getting rid of the overcapacity, they chose to go for the last pound of whale.

But these are all hypotheticals. When the blubbering has stopped, what it all boils down to is information and faith. We did not have the information to regulate whaling. And even when the information came dribbling in, the whalers chose to put their faith elsewhere. If they had started managing sooner, and done so rationally, they might not now be facing international hostility as they continue to go against the will of the people and the decisions of the IWC.

There will always be whaling. Whales represent too plentiful a bounty to ignore them completely. So long as the take is for local consumption only, whether by Eskimo hunters in the high Arctic or Caribbean fishermen in the tropics, it can do little harm. Some of the stocks, it is true, were so depleted by commercial whaling that they might not now be able to sustain even a limited subsistence hunt. But as a general rule local whaling is unlikely to deplete a population. It is only when the products of the whale become items of commercially oriented trade that whaling threatens the whales as species. That will be true as long as whales grow more slowly than money, which is likely to be for a very long time.

Sensible harvesting apparently has no future. It would, therefore, be nice if the whalers recognised this and abandoned their efforts sooner rather than later, but one cannot blame them for pursuing the final profitable whale. There are, of course, other arguments not to whale. One is that it is inhumane.

The standard weapon for killing a whale is the explosive harpoon. This has a barbed tip and is fired from a cannon mounted on the catcher boat. A steel cable runs from the boat to the harpoon head. Once inside the whale's blubber, spring-loaded barbs on the harpoon fly open to embed the head firmly in the whale. Screwed to the head is a bomb-point packed with explosives that are detonated on contact. The shrapnel and blast hasten the whale's death, which generally occurs through a combination of exhaustion and drowning. Some whales die within twenty minutes. Others may take twenty-four hours.

Minke whales are hunted with a non-explosive harpoon. Shrapnel fragments would damage an unacceptably high proportion of the valuable meat of these small whales. Minke whales therefore do not die as quickly as other whales. The IWC banned the use of non-explosive harpoons, but the countries that take the most Minkes, Norway and Japan, objected to that decision and continued to use so-called cold grenades.

Some whale supporters maintain that considerations of humanity offer sufficient reason not to kill whales. I cannot agree. No form of death is humane, and once you start to draw lines you are in terribly difficult terrain. An absolutist position —people should not kill animals—is at least consistent and defensible. But to say that we should kill fish but not whales is harder to justify. One ends up attempting to balance human need against animal suffering. Perhaps whales do have a greater capacity to suffer than, say, cattle. Perhaps people do need beef more than they need whale meat. If both statements are true then the humane argument may carry some weight, but I would not rely on it. After all, the working whaler's need for whale meat is considerably higher than anyone else's. So I do not think that the fact that killing whales may be cruel will ever stop anyone who is intent on killing whales.

There is an even more useless argument, which is simultaneously the most powerful and the most pathetic. Whales are magnificent. The largest creatures ever to inhabit the earth, they are supremely adapted to their environment. They are beautiful in their majesty. The notion that they should die to provide lubricating oil for tanks, or margarine, or fuel for lamps, or food for dogs and cats, or hoops for corsets and crinolines, or delicacies for the palate, is repugnant. But there are so many beautiful, awesome, unique creatures that we use all the time. What is so special about whales?

This, I confess, was my point of view. At the outset I really could not see what all the fuss was about. Either whales are a bounty available for exploitation, or they are not. If it turns out that they cannot be exploited in a sustainable fashion, well, that may be bad news for the whales in the short term, but such is the way of the world. Aesthetic concerns have nothing to do with it. Then I saw some whales in the wild.

I visited the Sea of Cortez, in Baja California, on a whale-watching cruise, the newest way of exploiting whales. I saw Grays, Minkes, Seis, Fins and Humpbacks, not to mention vast numbers of dolphins and porpoises. At one stage our boat was surrounded by a school of about twelve Blue whales, all feeding. They were spectacular to watch, their sleek blue-grey bodies easing through the water. When a whale dived down its tail left a fluke print of still, taut water on the choppy seas, a remarkable display of their strength and power.

I was standing on the bridge, and at one point I looked down and saw a whale's blowhole no more than four or five metres off our port bow. It breathed, with a sonorous resonating moan, and swam slowly on. I glanced back, and saw its tail fins on a level with our stern. The boat was 95 feet long. The whale was a few feet shorter.

Just ahead another whale crossed our path. As it did so a cloud issued behind it. Shimmering dots of red light glimmered in the sunlight, like laser beams against the deep blue of the water. The whales had been feeding on swimming crabs, known for their carotein pigments. Those pigments pass through the whale's digestive tract unchanged. Even the red shit of Blue whales is lovely.

None of this matters at all. A Blue whale may be a spectacular sight, enough to induce an unexpected feeling of cosmic oneness in a "hard-hearted" rationalist. But it is also 100 tonnes of meat, and just as no whaler is going to stop because it hurts the whale, no whaler is going to stop because someone else thinks his prey is wondrous.

Where does this leave us? Unless we start again from scratch, and construct a new whaling authority and a greatly diminished fleet, there is no hope of sustained whaling. That leaves unsustained whaling, or no whaling. Given their past history, and their current machinations, I do not think it is reasonable to expect the few remaining whalers to exercise any self-restraint. They are, I fear, likely to take the last whale they can. It may not be money alone that finally stops them. There may be greater concerns, like world standing and economic realities, but I think it would be foolish to assume any whales will be given up while there is still a hope that they can be taken.

There is, in fact, no reason to whale. The meat forms an insignificant part of the average Japanese diet. There are substitutes available for every product of the whale. Equally, there is no reason not to whale. The protein may be unimportant, but it is favoured. The other products do have their uses. In the end, it is a question of judgement.

We have the power to take the whales, although we do not

need to. We have the power to save the whales too, though we do not need to.

Why not save them?

Save them so the next generation has the privlege of seeing the wondrous creatures.

Acknowledgements and References

Many sources have informed my work on whales: other books; articles in the popular and scientific press; the newsletters of various organisations; unpublished papers; radio and television programmes; the meetings and staff of the International Whaling Commission. I have listed some of the more accessible sources below. I am also very grateful to all the actors in the whaling drama, many of whom took the time to explain their point of view to me. If I tried to thank them individually, I am sure I would forget some, so I hope they will accept this blanket expression of my gratitude.

Anon. (1978) "Endangered Species", *Time*, 10 July, 1978.

Anon. (1979) Report of the Sub-Committee on Protected Species and Aboriginal Whaling. IWC/31/4 Annex H.

Anon. (1982) A Summary of Experiments on Humane Catch of Whales, IWC/34/44.

Anon. (1984) Agreement to end Japanese whaling announced, United States Department of Commerce News. NOAA 84–75.

Anon. (1985) *Scientific Whalers? The history of whaling under special permits*. Greenpeace.

Anon. (1985) Report of the Sub-Committee on Protected Species and Aboriginal Whaling. IWC/37/4 Annex H.

Ballantyne, Robert Michael (1897) *The World of Ice, or the Whaling Cruise of "The Dolphin" and the Adventures of her Crew in the Polar Regions*. Nelson: London.

Barzdo, Jon (1981) *The Slaughter of the Whale*. RSPCA: Horsham.

Beddington, J. (1978) "Crisis for sperm whales", *Nature*, 276: 656.

Beddington, J. (1980) "How to count whales", *New Scientist*, 17 July, 1980, pp 194–196.

Bockstoce, J. (1981) "Man's exploitation of the western Arctic bowhead", FAO Fish Series (5), *Mammals in the Seas, 3*, 163–170.

Bockstoce, John (1980) "Battle of the Bowheads", *Natural History*, May 1980: 53–61.

Boeri, David (1983) *People of the Ice Whale: Eskimos, white men, and the whale*. E. P. Dutton: New York.

Brownwell, R. L., Jr and Ralls, K. (1986) "Potential for sperm competition in baleen whales", Report International Whaling Commission (Special Issue 8).

Bryden, M. M. and Harrison, Richard eds (1986) *Research on Dolphins*. Clarendon Press: Oxford.

Budker, Paul (1958) *Whales and Whaling*. Harrap: London.

Bullen, Frank T. (1899) *The Cruise of the Cachalot*. D. Appleton & Co.: London.

Carter, N. and Thornton, A. (1985) *Pirate Whaling 1985*. Society against Violation of the Environment (SAVE) International: Glasgow.

Chapman, D. G., de la Mare, W., Holt, S. J. and Payne, R. (1986) "Disregarding history". Statement issued 8 June 1986, Malmö, Sweden.

Cherfas, Jeremy (1979a) "Whale meat, again", *New Scientist*, 5 July, 1979: 2.

Cherfas, Jeremy (1979b) "The great white wash", *New Scientist*, 19 July, 1979: 175–177.

Cherfas, Jeremy (1980) "Among the sharks and whales", *New Scientist*, 31 July, 1980: 353–4.

Cherfas, Jeremy (1981) "Whaling: some ups but some downs", *New Scientist*, 30 July, 1981: 271–2.

Cherfas, Jeremy (1982) "Conservationists in a whale of a row", *New Scientist*, 4 February, 1982, p 295.

Cherfas, Jeremy (1984a) "Secret deal gives all clear for Japan's whalers", *New Scientist*, 15 November, 1984, p 4.

Cherfas, Jeremy (1984b) "New deal harpoons Whaling Commission", *New Scientist*, 6 December, 1984, p 4.

Cherfas, Jeremy (1985) "New moves may scupper whaling ban", *New Scientist*, 21 February, 1985, p 6.

Cherfas, Jeremy (1986) "What price whales?" *New Scientist*, 5 June, 1986: 36–40.

Cherfas, Jeremy (1987) "With harpoon and pocket calculator", *New Scientist*, 2 July 1987: 31–2.

Clark, C. W. (1976) *Mathematical Bioeconomics: the Optimal Management of Renewable Resources*. Wiley-Interscience: New York.

Clark, C. W. & Lamberson, R. (1982) *An Economic History and Analysis of Pelagic Whaling*. Marine Policy, April 1982: 103–20.

Comrie-Greig, J. (1986) "The South African Right whale census", 1985, *African Wildlife* 39: 222–6.

Darling, Jim (1984) "Source of the humpback's song", *Oceans* 17 (2): 3–11.

Day, David (1987) *The Whale War*. Routledge and Kegan Paul: London.

Donovan, G. P. ed. (1986) "Behaviour of whales in relation to management", International Whaling Commission: Cambridge.

Dow, George Francis (1985) *Whale Ships and Whaling*. Dover Publications: Mineola.

Dudley, N. and Gordon Clark, J. (1983) *Thin Ice*. Marine Action Centre: Cambridge.

ECO, "an occasional newspaper published by Friends of the Earth International and others at international meetings of importance to the global environmental community," is a major outlet for otherwise undocumented stories.

Ellis, Richard (1980) *The Book of Whales*. Knopf: New York.

Friends of the Earth (1978) *Whale manual '78*. Friends of the Earth: London.

Frizell, J., Plowden, C., and Thornton, A. (1980) *Outlaw Whalers 1980*. Greenpeace: San Francisco.

Gatenby, Greg ed. (1983) *Whales: a celebration*. Little, Brown: Boston.

Haley, Delphine ed. (1986) *Marine Mammals of Eastern North Pacific and Arctic Waters*. Pacific Search Press: Seattle.

Hardin, Garrett (1968) "The tragedy of the commons", *Science*, 162: 1243–8.

Harrison Matthews, L. (1978) *The Natural History of the Whale*. Columbia University Press: New York.

Holt, Sidney (1985) "The last pound of blubber", *BBC Wildlife*, January 1985: 20–1.

Holt, S. (1985) "Japanese Minke whaling in the Antarctic", International League for the Protection of Cetaceans Occasional Paper, 14 January, 1985.

Holt, S. (1985b) "Let's all go whaling", *The Ecologist*, 15 (3): 113–24.

Holt, Sidney (1985) "Whale mining, whale saving", *Marine Policy*, July 1985: 192–213.

Holt, S. (1986) "Mammals in the sea", *Ambio*, 15: 126–33.

"In defence of whaling: the Japanese speak out" (1980), Japan Whaling Association: Tokyo.

Jackson, Gordon (1978) *The British Whaling Trade*. A. & C. Black: London.

Jenkins, J. T. (1921) *A History of the Whale Fisheries*. Witherby: London.

Jones, Mary Lou, Swartz, Steven L. and Leatherwood, Stephen, eds (1984) *The Gray Whale*. Academic Press: London.

Kanwisher, John W. and Ridgway, Sam H. (1983), "The physiological ecology of whales and porpoises", *Scientific American*, June 1983: 111–118.

Kirkevold, Barbara C. and Lockard, Joan S., eds (1986) *Behavioral Biology of Killer Whales*. Alan R. Liss: New York.

Klinowska, Margaret (1986) "The Cetacean magnetic sense —evidence from strandings", in M. M. Bryden and Richard Harrison, eds, *Research on Dolphins*, Oxford University Press: Oxford. p 401.

Klinowska, Margaret (1987) "No through road for the misguided whale", *New Scientist*, 12 February, 1987: 46–8.

Leatherwood, Stephen and Reeves, Randall R. (1983) *The Sierra Club Handbook of Whales and Dolphins*. Sierra Club Books: San Francisco.

Mackintosh, N. A. (1965) *The Stocks of Whales*. Fishing News (Books): London.

Mark, Albert (1980) US Statement on Subsistence Whaling, 21 July, 1980, AM80–102W.

May, R. M., Beddington, J. R., Clark, C. W., Holt, S. J. and Laws, R. M. (1979) "Management of multispecies fisheries", *Science*, 205: 267–277.

Melville, Herman (1972) *Moby-Dick; or The Whale*. Penguin: Harmondsworth.

Montagu, Ashley and Lilly, John C. (1963) *The Dolphin in History*. William Andrews Clark Memorial Library: Los Angeles.

Murphy, Jamie (1986) "A deadly roundup at sea", *Time*, August 4, 1986: p 46.

Murphy, Robert Cushman (1947) *Logbook for Grace*. Macmillan: New York.

Olafsson, Arni (1986) Account of the pilot whale hunt in the Faroe islands. Danish Foreign Ministry: Copenhagen.

Ommanney, F. D. (1971) *Lost Leviathan*. Hutchinson: London.

Pain, Stephanie (1987) "Age miscalculations threaten bowhead whales", *New Scientist*, 5 March, 1987, p 21.

Payne, Roger (1982) "New light on the singing whales", *National Geographic*, 161 (4): 463–477.

Purves, P. E. and Pilleri, G. E. (1983) *Echolocation in Whales and Dolphins*. Academic Press: London.

Ridgway, Sam H. and Harrison, Sir Richard, eds (1985) *Handbook of Marine Mammals Volume 3: The sirenians and baleen whales*. Academic Press: London.

Scammon, Charles M. (1874) *The Marine Mammals of the North-western Coast of North America; together with an account of the whale-fishery*. Putnam's: New York.

Scheffer, Victor B. (1969) *The Year of the Whale*. Souvenir Press: London.

Scoresby, W. (1820) *An account of the Arctic regions with a history and description of the northern whale-fishery*. Constable: Edinburgh.

Small, George L. (1971) *The Blue Whale*. Columbia University Press: New York.

Swartz, Steven L. and Jones, Mary Lou (1984) "Mothers and calves", *Oceans*, 17 (2): 11–19.

Swinbanks, David (1987) "Japan disputes need for moratorium on whaling", *Nature*, 326: 732.

Taylor, Michael (1986a) "Stunning whales and deaf squids", *Nature*, 323: 298–9.

Taylor, Michael (1986b) "Sound strategies for survival", *New Scientist*, 30 October, 1986, 40–3.

Thornton, Allan and Gibson, Jennifer (1986) *Pilot Whaling in the Faroe Islands*. Crusade against Cruelty to Animals: London.

Tønnessen, J. N. and Johnsen, A. O. (1982) *The History of Modern Whaling*. C. Hurst: London.

Umealit: The Whale Hunters (1980) WGBH Educational Foundation, Boston.

Venables, Bernard (1968) *Baleia! the Whalers of the Azores*. The Bodley Head: London.

Walker, T. J. (1979) *Whale Primer*, Cabrillo Historical Association: San Francisco.

Watkins, W. A., Moore, K. E. and Tyack, P. (1985) "Sperm whale acoustic behaviors in the southeast Caribbean", *Cetology*, 49: 1–15.

Watson, Lyall (1980) *Sea Guide to Whales of the World: A Complete Guide to the World's Living Whales, Dolphins & Porpoises*. Hutchinson: London.

Weeks, W. F. and Weller, G. (1984) "Offshore oil in the Alaskan Arctic", *Science*, 225: 371–8.

"Whales vs Whalers: A Continuing Controversy" (1981) Animal Welfare Institute: Washington DC.

Whaling Review (1980) Japan Whaling Association: Tokyo.

Whipple, A. B. C. (1973) *Yankee Whalers in the South Seas*. Charles Tuttle: Rutland, Vt.

Winn, Lois King and Winn, Howard E. (1985) *Wings in the Sea: the Humpback Whale*. University Press of New England: Hanover, Vt.

Index